Edward Warren

An Epitome of Practical Surgery for Field and Hospital

Edward Warren

An Epitome of Practical Surgery for Field and Hospital

ISBN/EAN: 9783337163402

Printed in Europe, USA, Canada, Australia, Japan

Cover: Foto ©berggeist007 / pixelio.de

More available books at **www.hansebooks.com**

AN EPITOME
OF
PRACTICAL SURGERY,
FOR
FIELD AND HOSPITAL.

BY
EDWARD WARREN, M. D.
SURGEON GENERAL OF THE STATE OF NORTH CAROLINA, FORMERLY
PROFESSOR IN THE UNIVERSITY OF MARYLAND.

FRIST EDITION.

RICHMOND, VA.
WEST & JOHNSTON, 145, MAIN STREET.
1863.

TO

LAFAYETTE GUILD,

Surgeon C. S. A.

AND

JOSEPH P. LOGAN,

Surgeon P. A. C. S.

This work

IS MOST RESPECTFULLY AND AFFECTIONATELY

DEDICATED.

PREFACE.

Much experience in Field and Hospital, has convinced me of the necessity for the publication of a work on Surgery, more elementary, practical and concise in its character than has hitherto appeared. I have therefore, devoted myself to the preparation of this volume, as a *vade mecum* for the Surgeons of the Confederate service, with the view of supplying the desideratum which exists in this regard.

Little claim to originality, either as to principles or details, is advanced in these pages; but I have mainly endeavored to glean from fields of experience far richer and broader than my own, such views, facts, and deductions as are most worthy of diligent study, and faithful preservation. The ablest authors on the various subjects discussed, have been freely consulted, and the intelligent reader will have no difficulty in discovering to what extent I am indebted to them for the substance matter of this volume. Wherever an issue has been made with standard authorities, it has been from an honest conviction of the absolute necessity for such a course, and from an earnest desire to advance the best interests of Surgical science.

So far as the typographical execution of this book is concerned, I must urge in extenuation of its imperfections, that the best printers are in the service, and that those who remain behind are too young and inexperienced to do proper justice to any author. For this reason many errors will be found in this edition which shall be corrected in a subsequent one.

For the invaluable statistical information contained in the Appendix, I am indebted to the courtesy of Surgeon Samuel Preston Moore, Surgeon General of the Confederate States, under whose intelligent scrutiny and able direction it was carefully collected by Surgeon Francis Sorrell, Inspector of Hospitals for the City of Richmond.

Whatever the merits or defects of this unpretending work, it has been undertaken in a spirit of loyalty and humanity: and it is now issued with the hope of contributing something to a cause in which every sentiment of my bosom is most warmly enlisted.

Should it be the means of saving a single life, of alleviating a pang of pain, or of inspiring one professional brother with a braver heart and a steadier hand in the hour of trial, my proudest aspiration will be more than realized.

INDEX.

		PAGE.
Abscesses,		78
Accidents after Amputations,		116
Acupressure,		228
Alteratives,		78
Amputation, varieties of		81
"	primary,	81
"	secondary,	87
"	modes of,	104
"	of great toe,	137
"	of metatarsal joint	138
"	of metatarsal bones,	138
"	through tarsus,	139
"	at ankle joint,	140
"	of leg,	142
"	at knee joint,	144
"	of thigh,	146
"	at hip joint,	152
"	of fingers,	155
"	at wrist joint,	158
"	of fore arm,	159
"	at elbow joint,	160
"	of upper arm,	161
"	at shoulder joint,	162
Anæmia from loss of blood,		209
Antiphlogistic regimen,		62
Applications, cold and warm,		73
Arteries, structure of,		239
"	compression of,	218
"	ligation of,	235

	PAGE.
Arteria innominata,	266
Anterial sedatives,	57
Axillary artery,	268
Barton's fracture,	376
Blisters,	71
Blood, changes in,	13
" condition of,	213
Blood letting,	54
Bones, excision of,	101
" reproduction of,	203
" fractures of,	340
Brachial artery,	27
Brainard's perforator,	348
Cancer.	43
Causes of Hemorrhage,	211
Carpal bones, dislocation of,	323
" " fractures of,	379
Carotid artery,	247
Circular method of amputation,	104
Colle's fracture.	376
Comminuted fractures,	342
Complicated fractures,	342
Compound fractures,	342
" " of inferior maxillary,	368
" " of arm,	373
" " of hand,	378
" " of ribs,	366
" " of pelvis,	375
" " of thigh,	383
" " of leg,	388
" " of foot,	301
Compression of Arteries,	217
" of brain,	351
Concussion of brain,	351
Corpuscles, red,	13
" white,	13
Depletory remedies,	54

INDEX.

	PAGE
Depression of bone,	339
Disarticulation of fingers,	157
Dislocations,	295
" of lower jaw,	308
Dislocations of clavicle,	310
" of acromion,	312
" of shoulder joint,	313
" of ulna,	318
" of radius,	318
" of wrist,	322
" of thumb,	324
" of thigh,	326
" of patella,	333
" of tibia,	333
" of fibula,	336
" of astragalus,	336
" of calcaneum,	337
Effects of gun-shot wounds,	356
Erysipelas,	123
Extension and counter extension,	304
Fever,	26
Fissures of bone,	369
Flaps, length of,	109
Flap amputations,	105
Flexion of bones,	339
Fractures, classification of,	328
" of pelvis,	370
" of humerus,	370
" of ulna,	374
" of clavicle,	366
" of scapula,	653
" of cranial bones,	350
" of radius,	375
" of ulna and radius,	376
" of carpal bones,	379
" of fingers,	378
" of femur,	379

	PAGE.
Fractures of patella,	387
" of tibia,	380
" of fibula,	380
" of fibula and tibia,	370
" of bones of foot,	300
Gangrene,	189
Ginglymoid joints.	293
Hemorrhage,	210
Hemorrhagic fever,	208
Hernia cerebri,	361
Hospital gangrene,	120
Inclined plane,	382
Induration,	37
Inflammation,	13-70
Ligation of arteries,	226
" of arteria innominata,	246
" of common carotid,	247
" of exterior carotid,	250
" of thyroid.	252
" of lingual.	252
" of facial,	253
" of subclavian,	255
" of axillary,	268
" of brachial,	271
" of radial,	275
" of ulna,	278
" of common iliac,	356
" of external iliac,	264
" of internal iliac,	262
" of femoral	280
" of popliteal,	285
" of posterior tibial,	291
" of anterior, tibial,	287
" of dorsalis pedis,	389
" of peroneal,	289
Ligatures, mediate and immediate,	226
Litters,	345

	PAGE.
Litter Corps,	344
Lymph, absorption of,	36
Malar bone, fractures of	263
Median Basilic Vein,	274
Mercury,	57
Morphia—Endermic use of	305
Nervous Sedatives,	89
Opium,	59
Organization of Lymph,	43
Orbicular Joints,	298
Oval method of Amputation,	106
Palmar arch,	279
Points for ligatures,	235
" of stagnation,	14
Position in hemorrhage,	221
" " inflammation,	71
Pullies compound,	328
Pus,	40
Pyæmia,	127
Pyogenic membrane,	39
Redness,	18
Resections in general,	180
Resection of meta: carp: phal: articulation,	180
" " meta: carp. bones,	180
" " wrist joint,	180
" " radius,	181
" " ulna,	182
" " elbow joint,	183
" " shoulder joint,	185
" " clavicle,	190
" " scapula,	190
" " tarsus,	191
" " ankle joint,	192
" " knee joint,	192
" " hip joint,	199
Resolution,	36
Revulsives,	77

INDEX.

	PAGE.
Saline elements,	14
Seton,	228
Shock,	120
Smith's ant: splint,	386
Statistics,	392–400
Stumps conical,	119
" neuralgia of,	120
Structure of arteries,	239
Swelling,	20
Styptics,	236
Suppuration,	139
Tetanus,	131
Tourniquets,	217
Transformation of tissue,	45
Trephining,	169
Warm applications,	75
Water dressings,	74
Wounds of head,	356
" of face,	862
" of lungs,	368
" of arteries,	208
" of soft parts, generally.	357
" bones,	356
" entrance.	358
" exit,	358
Wounds from round balls,	357
" " conical balls,	358
" " swords,	352
Ulna,	313
Venesection in lung wounds,	868

ERRATA.

Page 19, 12th line, last word, read "inflamed."
" 58, 9th " next to last word, read "relieving."
" 132, 18th " after principle, read " nerve."
" 146, 13th " for " 60," read " 50 percent."
" 176, 31st " for " only," read " generally."
" 252, 14th " for " antrim," read " antrum."
" 285, 10th " for " abductor," read " adductor."

CHAPTER I.

INFLAMMATION.

DEFINITION.—Inflammation is a condition of altered nutrition in which a perversion of the Blood and Blood Vessels occurs, accompanied by increased vascularity, augmented sensibility, change in secretion, an exudation of Liquor Sanguinis, and a modification of structure and of function.

CHANGES WHICH TAKE PLACE IN THE BLOOD.—The blood becomes *thinner*, as was first established by Hewson.

The *Red Corpuscles* are increased in quantity in the early stages of inflammation, but are subsequently decreased as the disease advances. They also have a tendency to *cluster* together, by the cohesion of their flat surfaces.

The *White Corpuscles* are largely increased in number, and, by adhering to the walls of the vessel, tend to arrest the circulation.

The *Liquor Sanguinis*.—Andral and Gavarret have shown that Fibrin may be increased up as high as 6 per 1000—an augmentation which is manifestly due to the more rapid and complete metamorphosis which takes place in the tissues of the part. But by far the most remarkable and important fact which manifests itself in this regard, is the tendency to *effusion* which is developed as the disease advances. The Liquor Sanguinis escapes from the vessel and disseminates itself through the surrounding tissues, either, to be subsequently re-absorbed, to organize, or to break down into purulent matter. This exudation, according to Virchow, is the essential element in the inflammatory process, giving character to it, and furnishing the most reliable index as to its pathology and treatment.

The Saline Elements are somewhat below the normal standard, while the proportion of water is perceptibly increased.

The Coagulation of inflammatory blood takes place more slowly, while the coagulum is harder, and smaller, and the quantity of serum greater than under ordinary circumstances. The upper surface of the coagulum is covered with a layer of yellow fibrinous matter, known as the buffy coat, and depressed in its centre in the form of a cup.

The Buffy Coat, is regarded as an index and representative of the intensity of the inflammation, though the test is by no means infallible, in as much as the same phenomenon is manifested in Rheumatism, Pregnancy, and Plethora, in all their stages and conditions, without regard to the extent of the inflammatory process.

Points of Stagnation may be found, upon a close examination of the inflammed tissue, at which the blood current appears to ebb and flow, until it is finally and permanently arrested. This stagnation usually occurs in those capillaries which are not directly located between arteries and veins, and results from adhesion of the Red Corpuscles and the consequent blocking up of the vessels by the masses thus formed. It is at these points also, that the drawing away or exudation of Liquor Sanguinis usually commences,—facilitating the coalescence of the corpuscles, and indirectly contributing to the arrest of the blood current at the particular localities in question. Wherever this retardation of the circulation occurs, the white corpuscles may likewise be found in great quantities, either rolling slowly along the walls of the vessel or closely adhering to them.

CHANGES WHICH TAKE PLACE IN THE BLOOD VESSELS. —The arteries, capillaries and veins are usually contracted in the first instance, but are subsequently enlarged.

The Arteries leading to the part are especially dilated, while their coats are relaxed, so that the pulsations within them are stronger and more perceptible.

The vessels, in consequence of this dilatation, actually convey more blood to the inflammed part, than under ordinary circumstances, as has been repeatedly demonstrated.

In consequences of the expansion of the smaller arteries and capillaries, red corpuscles are more freely admitted, so that the part becomes red, as if from the development of new vessels.

The arteries are not only dilated, but become elongated, tortuous, and waving—increasing in length as well as in circumference, while small branches project from their walls, and fusiform dilatations of the whole diameter frequently present themselves.

The distention of the arteries and capillaries *before* the point of obstruction, induces increased effusion of serum, lymph, and pus.

The veins *beyond* the point of obstruction are empty; and, hence, there is increased absorption with softening &c.

The circulation *at* the point of obstruction is arrested, so that there is a reduction or abolition of the vital properties; and consequently, either gangrene, ulceration or suppuration is developed.

There is also increased circulation of the blood *around* the point of obstruction, causing exaltation of the vital properties; and, hence, spasm, pain, sympathetic irritations, increased secretion &c., are produced.

CHANGES WHICH ARE INDUCED IN THE SYSTEM AT LARGE.—The excitement may extend to the heart and arteries, causing inflammatory fever.

The whole mass of blood may undergo alterations by increase of fibrin, by diminution of the secretions, and by the retention in the circulation of their elements.

Exhaustion ensues after excessive excitement, the effusion of serum or the formation and escape of pus.

Depression, with partial irritation, not unfrequently supervenes in consequence of the presence of pus in the blood. Though the pus globules cannot be absorbed into the blood by reason of their size, in their normal state, yet certain modifications take place in them, under some circumstances, which do admit of their being taken into the circulation,—causing the development of peculiar symptoms and the induction of fatal consequences.

CAUSES OF INFLAMMATION.—The causes of inflammation may be divided into *predisposing* and *exciting*.

Predisposing causes act both locally and generally. The general or constitutional predisposing causes are plethora; excess in food, or bodily exertion; exposure to miasmatic influences; disorders of the liver, skin and kidneys; great mental emotion; over stimulation; vascular and nervous depression &c.

The local predisposing causes are, excessive use of the part; previous injury or disease; delicacy of organization &c.

The exciting causes are, mechanical injury; chemical agencies; morbid poisons; and certain imponderable agencies, as heat, cold, galvanism &c.

The causes of inflammation may produce their legitimate results either directly—that is to say, by irritating and inflaming the part with which they are in contact—or indirectly through the agency or instrumentality of *nervous reflex action*, as when cold applied to the feet causes inflammation of the lungs, bowels, or peritoneum.

The causes of inflammation are common or specific,—the former being of constant occurrence, and affecting all constitutions equally,—the latter being peculiar in their origin, action and effects upon the human economy.

SYMPTOMS OF INFLAMMATION.—The symptoms or signs by which inflammation is distinguished are local and general,—that is to say, connect themselves both with the part affected and with the system at large.

Local Symptoms.—The symptoms of inflammation which connect themselves with the part affected, are redness, pain, heat, swelling, alteration of structure and changes in function.

Redness—This is owing to the primary production of a greater number of red corpuscles than usual; to the enlargement of the vessel, permitting red corpuscles to circulate more extensively through them; and to the presence in the part of an unusual amount of blood.

Pain.—This is due to the tension of the nervous filaments directly involved; to the greater irritability of the whole nervous mass; and to the augmented susceptibilities of the sensorium. There are different varieties of pain. Thus, it is dull, obtuse, heavy or aching in congestions, and chronic inflammations, or in acute inflammations of parenchymatous, organs: it is gnawing or lacerating in rheumatism, gout, and periostitis: it is lancinating in scirrhus or in inflammations of the nerves: it is twisting, griping or spasmodic in dysentery, ileus, gastralgia, and obstruction of the intestines: it is burning as in cutitis, and erysipelas: it is sharp and cutting in inflammations of serous membranes: and it is oppressing in inflammations of the stomach, testicles, liver, and kidneys.

Pain is not an invariable concomitant of inflammation. Thus it is absent when inflammation only ends in adhesion;—when the inflammatory action is indolent, as in scrofula; when both the mental and physical susceptibilities have been destroyed, as by the abuse of spirituous liquors, opium and tobacco,—the exhibition of chloroform, or the existence of that peculiar morbid condition which is denominated insanity; when the nervous centres have lost their normal irritability or responsive power, in consequence of the absorption of some "blood poison," or the retention of the elements of the bile, urine, &c.; and, when the connexion between the brain and the affected part is destroyed, as by the destruction of the nervous filaments serving as their bond of union.

Heat.—The amount of heat in an inflammed part, is never so great as the patient supposes, though it has been established by the experiments of Becquerel, and Breschet, that Celsus and Hunter were correct in regarding elevation of temperature as a characteristic of the inflammatory process. The temperature of the foci of inflammation is to be regarded as the expression of several distinct sources of heat, viz:

1. From the blood which accumulates in an unusual quantity about the centre of irritation.

2. From the increased metamorphosis of tissue which takes place in consequence of this accumulation of blood, and the attendant superabundance of those elements whereby the structures are renewed.

3. From the more active metamorphosis of tissue which is induced by specific changes in the nervous *status* of the part.

The blood is *warmer* than the subjacent tissues, and hence, there must be more *heat* at those points where this fluid accumulates. Again, as it is now placed beyond question, that the source of the normal animal temperature is to be found in the chemical development of heat attendant on nutrient changes continually occurring in the tissues, it follows that, where there is an elevation of temperature, there must also be increased metamorphosis. Now, this increased metamorphosis becomes a matter of necessity when an unusual supply of pabulum is presented, as is the case where blood accumulates in tissues which have at once an appetite for it, and the power of appropriating it according to their necessities.

And lastly, the experiments of Bernard and Sequard, have clearly established, that this appetite of the tissues, or in other words, their formative power, or metamorphic capability, can be increased or diminished according to the amount of nervous influence supplied to them. It follows therefore, that when there is an excess of pabulum—as must occur when the circulation is more rapid than usual, or there is an increase of blood in the part from any cause, and such a concurrent change in its nervous condition takes place as tends to stimulate its nutritrive power,—there must be a more rapid and complete metamorphosis, and a corresponding elevation of temperature.

Swelling.—This is caused at first by the increased quantity of blood, and subsequently by the effusion of lymph, the pouring out of serum, or the formation of pus. The more dense the texture, the less there is of swelling, and *vice versa*.

Alteration of Function.—Each tissue and every

organ has a certain part to perform in the economy, which is its contribution to the completeness and perfection of the organism. This is known as the *function* of the part. Thus the *function* of muscular tissue is to contract, and of glands to secrete. Now, a given tissue requires two things particularly, in order to secure the proper performance of its appropriate function, viz: the distribution to it of a certain amount—neither too much nor too little—of nervous influence; and the preservation of its structures in their normal condition. Inflammation, as previously shown, not only changes the *nervous status* of the part, but constitutes *per se* such a veritable perversion of its nutrition, as speedily induces a positive modification of its structure.

It thus becomes plain, that inflammation, must, as a matter of necessity, materially interfered with the *function* of the part in which it has been produced, while all experience confirms the truth of this deduction.

It is in this way that alteration ot secretion ensues. Thus, secretion is usually *diminished* at the commencement of inflammation, *suspended* when it is at its acme, and *increased* at its close, if health be the termination. In the same manner, secretions may change their characters chemically, or become mixed with the products of inflammation, as blood, serum, epithelial cells, tube cast, lymph and pus.

Alterations in Structure.—These take place in consequence of the alteration in the nutrition of the part. The various tissues of the body are differently affected by the inflammatory process, as will be shown hereafter, but there are certain changes common to all of them, which may be mentioned here. The *weight* is usually increased, unless apoplexy be produced: hardness is diminished,—that is, there is less of cohesion in the part, because of the effusions which infiltrate its tissues. In chronic inflammations the opposite of this is frequently the case, inasmuch as the effused lymph organizes, or the whole limb may become shrunken: Transparency is destroyed. Polish is impaired materially: and alterations may take place in all the physical properties pertaining to the tissue.

General or Constitutional Symptoms.—The most promenent and important of these is fever. Fever and inflammation are processes that many confound with each other, though they are really distinct. They may alternate or be intercurrent; and on the other hand, their characters and phenomena may be so blended as to render it a matter of impossibility to draw a line of demarcation between them, and even to necessitate the use of a mixed term to define the resulting condition. It is in this way that the expression Inflammatory Fever has obtained a place in the vocabulary of medicine; and yet, whatever may be the analogy between them, or however undoubted the fact of their simultaneous existance at certain times, it is impossible to deny that they differ in their essential nature, and that they are totally distinct processes.

Points of Resemblance between Fever and Inflammation.—The following characteristics distinguish both of them:

1. An elevation of the animal temperature, such as can be distinguished and measured by the thermometer.

2. An acceleration of tissue metamorphosis of a decided and appreciable character.

3. An increased rapidity of the circulation and definite changes in the nervous system, as have already been referred to, and as will be more fully explained hereafter.

Points of Difference between Fever and Inflammation.

1. Inflammation is usually of local origin, whilst fever is generally of systemic origin, and in its course involves the whole organism.

2. In Inflammation, the attending heat, acceleration of metamorphosis, excitement of circulation and change of nervous *status* is localized; while in fever these conditions are produced generally and simultaneously throughout the system.

3. In Inflamation metamorphosis is induced in the tissues even to the extent of their disorganization. In Fever, the nutrient local changes, though accompanied by interstitial absorption, progress, both in tissues and organs, without material interference with their functions.

4. Inflammation usually results as the effect of some mechanical, or chemical cause, acting upon the animal structures, and can be produced at will. Fever, on the other hand, is produced by causes which can be neither explained nor controlled.

5. The Inflammatory process can be checked, controlled or modified by the employment of proper therapeutical agents; while, of most fevers, it may be asserted, that they are self-limited, and that any attempt to cut them short must result in failure as a matter of necessity.

In this connexion, Lyons,* uses the following appropriate and significant language: "While I believe it may be said with truth that we can *cure* many Inflammations by the intervention of art, the same cannot be affirmed of Fevers. In Fevers the highest efforts of our art, the most delicate care, the most refined skill, the most nice appreciation and adaptation of means to ends which we can command, must be all directed to watching, supporting, maintaining, and it may be stimulating the system till the fever-storm shall have passed over it."

Circumstances under which fever is not readily produced.

1. When the Inflammatory process limits itself simply to the repair of tissues, Fever is not one of its attending phenomena.

2. When it is circumscribed, that is, when, but a small portion of the animal structure is involved, Fever is not ordinarily developed.

3. When it occurs in tissues of an inferior degree of vital organization, the system does not respond to the local impression, and that reaction,

---

* A Treatise on Fever &c., by Robert D. Lyons, K. C. C. D. D. Blanchard & Lea, Philadelphia, 1861. To this admirable work, we are indebted to many for the above views.

which we denominate febrile excitement, is not produced.

Thus, an inflammation of the skin, cellular tissue &c., does not produce fever so readily as inflammation of the parenchyma of the lungs, of the pleura, or of the synovial membranes.

4. When it occurs in persons whose constitutions are neither above the standard of health, nor below—neither *plethoric* nor *anæmic*—, fever is not readily produced.

Circumstances under which Fever is readily produced.

1. When the Inflammatory process assumes a greater degree of violence than is necessary for the repair of tissues, and threatens the disorganization of the part.

2. When it involves a considerable portion of the animal structures.

3. When it affects tissues which possess a high degree of organization. Instances in explanation of this point were given under the last head, though if additional proof be wanting, reference can be made to the facility with which Fever is developed in connexion with Inflammation of the delicate coats of the eye, of the nerves and of the internal tunics of the blood vessels.

4. When it attacks parts which have numerous and important nervous connexions with the system at large. In this way fever is developed either *directly*, or *indirectly* by what is known as *nervous reflex action*. Thus Inflammations of the brain, spinal cord, and stomach readily and rapidly produce an impression upon the whole system, which expresses itself in febrile-excitment.

5. When it occurs in persons whose constitutions possess an unusual degree of susceptibility to local impressions and general influences of a morbid character.

6. When it is developed in those whose nervous systems are particularly irritable because of the existence of *plethora,* or of *anæmia,* though in the one *instance* the fever assumes a *sthenic* character whilst in the other, it is of a *low grade.*

7. When it exists in connexion with an epidemic of fever, the development of malarial poison, or those debilitating influences which are the prolific sources of typhoids, and typhus, such as infest crowded camps, ill-ventilated Hospitals, and the confined Burden Cars in which soldiers are so frequently transported.

Definition of Fever.— It is a matter of the first importance to understand the exact meaning of the word Fever, to comprehend the precise pathological conditions which are included in and expressed by that most significant term.

From the days of Celsus to the present time, the Profession has sought eagerly for a proper definition of Fever; but it is generally agreed, that Cullin's description embodies the most correct enumeration of its essential phenomena. It is as follows: "after a preliminary stage of languor, weakness, and defective appetite, there occur acceleration of the pulse, increased heat, great debility of the limbs, and disturbance of most of the functions, without primary local disease."

Phenomena of Fever.—Essential Phenomena.— Galen long since declared that the *essence* of fever consists in a *calor præter naturam*, and the most recondite researches and scientific analyses have succeeded in discovering no element that is more characteristic, constant and important than the *elevation of temperature* which invariably accompanies the febrile paroxysm. That there is such an elevation has been decided by the experiments of De Haen, who found, that even in the *algid states of fever*, there was, in the internal parts, a manifest increase of temperature, in some cases to the extent of 2°, 3° and even 4°, and that the slightest febrile conditions are attended with an increase of heat, which is likewise in some instances the only observable phenomenon whatever.

The chief source of the increased temperature in Fever is to be sought in an exaggeration of those causes which operate in the production of heat in the normal state of the system. It is now universally admitted, that the source of the normal temperature is to be found in the chemical development of heat, which results from the nutrient processes invariably occurring in the various structures of the organism. It follows therefore, that the *elevation of temperature which characterizes the febrile condition, is the result and the exponent of an accelerated metamorphosis in the tissues*. It must be remarked, in this connexion, that there is not only an increased consumption of the natural pabulum which the blood supplies to tissues, but that the actual constituent elements of the body themselves are appropriated and removed by the increased metamorphic activity induced in the structures generally. Thus the fluids, the muscles, the adipose tissue, the glands and even the bones themselves waste away during the progress of a febrile attack, particularly if it be of long duration, or of great intensity.

As the normal nutrition of the tissues bears a direct ratio to the amount of blood distributed to them,—since it is the source of their pabulum, it follows that the accelerated metamorphosis which characterizes the Febrile paroxysm, must be accompanied by an increased activity of the circulation. It is well known that however induced, an augmentation of the force and the rapidity of the circulation presents itself among the earliest concomitants of a large majority of febrile attacks. So invariable is this association, in fact, that alterations in the Pulse are universally regarded as an essential element of that peculiar condition which we denominate fever.

Metamorphosis, though depending to a great degree upon the amount of pabulum supplied by an increased circulation, or an accumulation of blood from any cause, is also, to a certain extent, influenced and controlled by the nervous system, since, as before remarked, *it* has the power of increasing the appetite of the ultimate elements, and of thus inducing a larger consumption of those materials upon which they feed.

It has been shown by Bernard, Sequard,— Weber, Virchow and others, that the nervous system exercises a direct and most potent control over the circulation. Thus, Bernard has demonstrated that the section of the sympathetic nerve in the neck is followed by a rapid increase of temperature in the corresponding side. Brown Sequard has cut the sympathetic filaments distributed to the ear of a Rabbit, and found, that there was not only an increase of temperature in it, but that the blood was warmer on *leaving*, than when it *entered* the part. Weber has shown, that irritation of the Vagi nerves causes an arrest of the heart's action; and it has been known for a long period, that after section of these nerves, an immediate and decided acceleration of the pulse takes place. Similar experiments have been made by Ludwig, Valkman, Fowelin, and Traube, and with like results; whilst Virchow has investigated the subject still farther, and with such success as to induce him to build upon the facts eliminated, the whole superstructure of his febrile pathology. For these reasons, it is now regarded as an established fact, that certain parts of the nervous system preside over the general and local circulations, and that all changes in them, depend upon and represent certain complimentary and precedent alterations in the nervous *status* of the organism. Virchow, who may be regarded as the great pathological pioneer of the 19th century, believes that these abberations affect primarily the *regulator* or *moderator* functions of the nerves, and that the nerves which play this important part in the economy are the Vagi and Sympathetic, having, in all probability, their centre, especially the former, in the Medulla Oblongata.

The essential phenomena of Fever may therefore be thus summed up:
1. Increased heat, produced by—
2. Increased metamorphosis, produced by—
3. Acceleration of circulation, produced by—
4. An irritation of the regulator nerves, especially the Sympathetic and Vagi, whose centres are in the Medulla Oblongata.

Non essential phenomena of Fever.—Fever may be accompanied by *pain* especially of the the head and loins; a sense of *heaviness* or *general lassitude;* deficiency *of either secretion,* or of all of them; *dryness of skin; thirst; nausia; scanty and high colored urine; delirium; constipation; jactation;* &c. Some one of these symptoms is always present in connexion with inflammatory action, but they constantly vary, and, on that account may be regarded as non essential phenomena.

Varieties of Fever.—Fevers may be divided, with reference to the causes producing them, into two great varieties, viz: *Idiopathic* and *Symptomatic.*

1. Idiopathic Fevers.—These are produced by causes of an inappreciable character, either developed within or without the organism, and acting upon the nervous system *directly* or *indirectly* through the agency of the blood. Typhoid, and Typhus are types of this class of Fever.

2. Symptomatic Fevers are produced by some injury or disease of a particular portion of the organism. They are, in fact, nothing more or less than the system's response to an impression made by a disturbing agency, upon some one of its parts—the general manifestation of a special

pathological disturbance. It is with Fevers of this discription that the Surgeon has specially to deal, and they must therefore be particularly considered in this connexion.

The essential elements of all fevers are identical, while their non-essential phenomena constantly vary. Heat: inceased metamorphosis; acceleration of the circulation; and nervous disturbance are the invariable phenomena which distinguish and characterize febrile action. Fever, then, regarded as a pathological entity—a distinct unit, made up of the peculiar morbid conditions just mentioned—is always the same so far as its essential nature is concerned. It is true that the degree of heat, the extent of the metamorphosis, the rapidity of the circulation, and the amount of nervous disturbance are exceedingly variable; but it is equally certain that the mode, order and history of their development are precisely the same under every variety of circumstances. It is therefore a misnomer to denominate fever *per se* as inflammatory, irritative, &c.; and, hence, the usual classification adopted by writers on this subject, is manifestly unphilosophical because it has no foundation in positive pathological fact.

Fever, however, may associate itself with the Inflammation of a healthy system, or with the Inflammation of a debilitated, impoverished cachectic system.

The *first* is known as Pyrexia, or true Surgical fever, and is of a *sthenic* character. Its *symptoms* are a hot and dry *skin*; a full, bounding and frequent *pulse*; the diminution or arrest of the *secretions; acidity and high color* of the *urine;* constipation of the *bowels;* coating of the *tongue* with a white fur; *thirst;* languor, heat and pain of *head.* A disposition is always manifested in this connexion, to *remit* or *intermit*, or in other words, the fever is not of a continuous character.

Its *abatement* is followed by the subsidence of all the symptoms mentioned above:—by a free perspiration;—by abundant discharge of urine abounding in *lithates;*—by a natural movement of the bowels, or it may be diarrhœa;—by cleansing of the tongue, abatement in the frequency and force of the pulse;—by subsidence of thirst and a general feeling of relief on the part of the patient.

The *second* which is of a decided *asthenic* character, presents itself under three forms, viz: *Typhoid Fever, Irritative* or *Nervous Fever,* and *Hectic Fever.*

The true Asthenic or Typhoid Fever occurs principally in persons whose constitutions are enervated by exposure, privation, irregularity of life, grief, or long residence in a vitiated atmosphere.

Symptoms.—The period of depression is marked and much prolonged. The reaction is not of a very active character: there is a disposition to heaviness, stupor, and delirium; the pulse is feeble but quick and frequent; the skin is sometimes moderately *hot,* then again is particularly *dry* and *burning*, and occasionally covered with an abundant *perspiration;* the cheeks are *flushed*, and the eyes bright and starring, while the tongue is *red. dry* and sometimes *cracked* in its centre.

The *abatement* of the fever is characterized by a gradual disappearance of all the symptoms; but the patient remains weak and debilitated for months, and the return to health is invariably slow and uncertain.

Should the disease take an unfavorable turn, the pulse grows more feeble and frequent, the tongue *dryer* and more *cracked*, the skin *cold* and mottled; while hiccup, subsultus, dyspnœa or coma comes on and death closes the scene by claiming its victim.

There is always a tendency to visceral complications in connexion with this affection, which not unfrequently decide the fate of the patient. The fever is usually *continuous* and pathologists locate the especial seat of the disease in the Sympathetic system.

Irritative Fever is a variety of the asthenic form though not of so specific a type as the last. The nervous system is especially concerned; and the affection presents itself in connexion with the systems of those whose mental powers have been over taxed, or whose vital energies have been destroyed by excessive venery, indulgence in drink, constant intellectual labor, &c.

The *symptoms* which distinguish Irritative fever are a dry and red tongue; a sharp, small, but frequent *pulse*; *subsultus*; *restlessness* and *delirium*, which soon give place to signs of debility, with coma and cerebral irritation, sudden *exacerbations*, unequal and irregular *remissions*; rapid and important *changes* are also frequent concomitants of this form of disease.

Hectic Fever, is also a variety of the *asthenic* form, and generally presents itself in conjunction with some organic, serious disease, excessive discharge of any secretion, but more particularly with the formation of *abcesses* and the production of pus. Emaciation; debility; clear and red tongue; disposition to diarrhœa and profuse perspiration; a frequent and small *pulse;* slight chills followed by burning of the hands and feet, with a circumscribed flush upon the cheek, indicating derangement of the capillary circulation, are the symptoms which characterize this form of fever.

Hectic is but too frequently the harbinger of a speedy death; and yet, it is really astonishing to observe with what rapidity and completeness many patients recover even after the development of its most characteristic and unfavorable symptomps.

It not unfrequently has the effect also of producing an exhileration of the spirits,—elevating them to such an extent as to preclude all fear of the fatal catastrophy of which it is the sad precursor.

TERMINATIONS OF INFLAMMATION.—Inflammation may terminate either in the *repair* of the part; in its *return to health;* in the *modification of its function and structure;* or in *its death.*

Repair of the part.—A part whose continuity has been broken or destroyed may be repaired, after the development of Inflammation, either by the immediate organization of the Effused Lymph, or by its more slow and gradual conversion into a structure identical with that of the subjacent tissues or similar to it.

When the repair is immediate, it is called union by the "First Intention," and when more tardy—

being accompanied by the formation of healthy pus, granulation, &c., it is denominated union by the " Second Intention."

Restoration of the Part to health.—Inflammation may be developed in a part, which has suffered no solution of continuity, under the influence either of some Local or General cause, and, after the manifestation of all the characteristic symptoms, of that process, leave it in its original condition. This is accomplished by the reabsorption of the effused Plasma, either in its nascent state, or after it has been changed into blastema and fibro-cellular tissue. The absorption of the Lymph in its liquid state is denominated Resolution, and is the most favorable termination or effect of Inflammation. Nature, in many cases, labors to make way with effused Lymph in such a manner as will prove least injurious to the surrounding parts as well as to the organism; and, hence, the work of reabsorption is commenced, under its watchful and intelligent guidance, to be perfected or not according to the circumstances of the case. Each petholigical step is then carefully and successfully retraced. The attraction between the Globules and the walls of the vessel, loses its intensity; the *stasis* of Blood disappears; the Heat, Pain and Swelling abate; and the part assumes its normal *status*, both as regards function and organization. It sometimes happens, however, that all of these steps are taken suddenly and simultaneously, or occur so rapidly as to be inappreciable. This is styled *Delitescence*.

Metastasis is the sudden translation of Inflammation from one point to another. This, in a majority of cases, may be regarded as a phenomenon of Nervous Reflex Action—a principle which plays a most important role both in the Physiological and Pathological processes of the organism.

Resolution is the natural, legitimate and most favorable conclusion of the Inflammatory process—a result towards which the efforts of the Practitioner should be invariably directed as the most effectual method of preventing disastrous consequences.

The Absorption of Lymph after its conversion into blastema and fibro-cellular tissue, does not occur to any considerable extent during the height of the inflammation by which it has been produced. There must always be a marked reduction of the morbid action before the absorbent vessels can be forced to take hold of it; but when this point has been once reached the process often goes on with great rapidity. When the Lymph has become completely organized, absorption is, of course, still more difficult, and not unfrequently impossible.

It is more than probable that Lymph even in a liquid state, is not absorbed until it has been dissolved in the fluids of the affected parts, when it is brought more readily under the influence of the absorbent vessels.

Modification of the structure and functions of the Part. Inflammation may also leave the Part modified both as regards function and structure. This modification is due to the influence of certain products of the Inflammatory Process, which should be briefly considered, in connexion with this mode of development, and the nature of the effects produced by them.

The effects or products of Inflammatory Action, which play this important part in the economy are: Induration, Hypertrophy, Atrophy, Effusion of Serum, Formation of Pus, Organization of the effused Lymph, or Transformation of Tissue.

Induration. When the effused lymph is not absorbed it organizes, either forming a sort of internal *cicatrix* which is harder than the surrounding tissues or increasing the density of the part by augmenting the amount of plastic material within it.

Softening. This results either from the infiltration of effused liquids, or disintegration of the substance of the textures themselves, by which their consistence is diminished.

Hypertrophy. It has been previously shown that the Inflammatory Process not only increases the amount of Blood—the pabulum sent to a given tissue—but also stimulates the appetite of the part, so as to render its nutrition more active and complete. It thus happens, not unfrequently, that tissues, and whole organs are permanently enlarged, as a consequence of Inflammation.—Hypertrophy is essentially a local disease.

Atrophy. Though atrophy is the opposite of Hypertrophy it is not an unusal effect of Inflammation. Nutrition is made up of two elements, which though entirely distinct, the one from the other, are absolutely essential to the perfection of the process. Cell destruction as well as Cell elaboration—the breaking down and the building up of tissue, occur simultainously throughout the whole organism. The term metamorphosis includes both of these processes; and in the normal condition of the system presupposes a perfect equilibrium between them. Under the influence of Inflammation this equilibrium is lost, so that cells may be too readily produced, or too rapidly destroyed. In the one instance Hypertrophy is produced and in the other Atrophy is the result.

Effusion of Serum.—Congestion or the accumulation of Blood in the part affected, constitutes one of the distinguishing features of Inflammation. It is, in fact, an essential element of that process. The vessels thus become filled with an unusual quantity of the circulatory fluid, which distends their coats, and facilitates the pouring out, or the *exosmosis* of the watery portion of the Blood into subjacent cavities or neighbouring tissues. It is in this way that fluxes are produced and dropsies occur, materially altering the structure of tissues and organs, and interfering with their peculiar functions. All the tissues do not present the same tendency to the effusion of serum in connexion with Inflammatory action. The structures which supply it in greatest abundance are the *cellular* and *serous*, the secernent vessels of which are extremely active even when the disease itself is comparatively mild. The mucous membrane of the alimentary canal, particularly that of the Colon and Rectum is frequently the source of large effusions of serum, as is seen in diarrhœa and Cholera Infantum.

The appearance of the serum is usually limpid,

though it may be changed by admixture with the secretions, Lymph or Pus. The effusion of Serum is always a phenomenon of *Osmosis*, while it is controlled by the laws which govern that process, and is dependent upon that principle alike for its production and its cure.

Suppuration or the formation of Pus. The idea was long entertained that Pus was a veritable secretion, poured out from the vessels under certain abnormal circumstances, and subject to all the laws which control the products of secerning organs generally. The researches and arguments or Gulliver, Mandt and Addison have demonstrated the incorrectness of this opinion ; and it is now generally agreed among Pathologists, that Pus Corpuscles are modifications of the Exudation Cells, and that suppuration is nothing more nor less than the breaking down or degeneration of the Lymph poured out in connexion with the inflammatory process.

When Lymph is not converted into tissue, or false membranes—because of the blight impressed upon the formative power of the contiguous structures by the Inflammatory action—or fails to organize even into cacoplastic products, a peculiar depreciation takes place in it whereby the corpuscles of the Plastic mass are transformed into Pus Cells, the Blastema degenerates into *liquor puris*, and Purulent matter takes the place of the more highly organized effusion.

When Pus is formed upon a free surface, it is styled a *Purulent secretion*; and when elaborated within the structure of a part, it is called an *Abscess*.

Nature usually makes an effort to retain the Pus thus formed within limited bounds, by depositing around it an external boundary of consolidated Lymph, known as the *Pyogenic Membrane*. This does not secrete Pus as was supposed by Delpech and many others, but is simply the boundary line between the abnormal product and the intact tissues. When this Membrane is absent, it may be regarded as indicative of a want of tone in the system, and as such furnishes a valuable hint to the Surgeon as regards prognosis and treatment. In the above remarks concerning this Pyogenic Membrane, the production of Pus in connexion with abscesses, is only referred to. This fluid is elaborated along the track of wounds extending through tissues of all grades and varieties, with so much rapidity and in such large quanties, as to preclude, even in the most vigorous constitutions, the formation of a protecting membrane, and is, hence, found diffused, in many instances, throughout the subjacent structures.

When Pus is opaque, thick, smooth, slightly glutinous, of a yellowish white color, with a greenish tinge, a faint odour and a alkaline reaction, it is said to be *healthy* or *laudable;* when mixed and tinged with blood it is *sanious;* when thin watery and acrid, *ichorous;* when it contains *cheesy* looking flakes, *curdy;* and when diluted with mucus or serum, *muco-pus or suro-pus.*

It consists, when laudable, of corpuscles, floating in a homogeneous fluid, styled "liquor puris." These corpuscles are modifications of the exudation-cells, and are composed of a semi-transparent

TERMINATIONS.

cell-wall, with two or three nuclei, of large quantities of granular matter, of particles of fibrin, and of disintegrated exudation cells. A multitude of changes, however, may occur in it, altering its composition, and changing its character, which can be detected by the microscope. When the suppurative process has once been set up, it may continue for an indefinite period, becoming, as it were, the fixed secretion of the part. From mucous membranes particularly, it has been known to last for years.

The symptoms which indicate that Pus is *about* to be formed, are; a more throbbing pain, a greater swelling and tension of the part, and a red, glazed and shining appearance of the skin, though it is sometimes elaborated without the development of any antecedent local sign.

The symptoms which indicate that Pus *has been* formed are: the disappearance of the ordinary signs of inflammation; the occurrence of chills or rigors; alternations of heat and cold; abatement of the intensity of the fever, and its assumption, in some instances, of an intermittent character; softening and perhaps quickening of the pulse; and fluctuation in the part, with enlargement also when the fluid is diffused throughout its tissues.

The symptoms which indicate that Pus is escaping from the system in too great a quantity, are: emaciation and loss of strength, a quick, small and compressible pulse; a coated and dry tongue with red tips and edges; flushed cheeks; dilated pupils; profuse sweating; copious purging; large discharges of urine, filled with red deposits; great debility; hypocratic countenance; husky voice; insomnia, &c. There is usually an exacerbation towards evening, and the actions upon the bowels, skin and kidneys alternate with each other, until the patient dies from sheer exhaustion.

The tendency to suppuration is increased by the following circumstances, viz:

1. Peculiar conditions of the Patient's system. Thus, in conditions of debility from any cause which diminishes the vital powers, as bad food, impure air, cachectic states of the organism, scrofula, &c.

2. Specific character of the Inflammatory process. Thus, in Gonorrhœa and Purulent Opthalmia, Pus is more readily eliminated than under ordinary circumstances.

3. Locality of the Inflammation. Mucous membranes more readily suppurate than serons, &c.; cellular tissues more rapidly than muscular, &c.; Inflammatory surfaces when exposed to atmospheric air supurate more readily and freely than others.

4. The state of the part affected. All parts of the system are not invariably in the same condition of health. Thus the nerves running to a particular part may have been divided by some previous accident, or some affection peculiar or confined to it may have lowered the tone of its vital powers, &c. In this way Inflammations which some portions of the body would readily resist, terminate elsewhere in suppuration.

The Plastic matter thus destroyed is the food of

the tissues involved in the Inflammatory action—the pabulum upon which they depend for the preservation of such properties as are essential to the integrity of their structure and the perfection of their functions.

Again, the purulent fluid by desseminating itself throughout the tissues, or by pressing upon them, so changes the normal *status* of the part as to disqualify it, either partially or completely, for the performance of its proper offices.

Of the fatal consequences which connect themselves with the presence of Pus in the blood, it is unnecessary to speak in detail here, in as much as they will be more fully discussed in another connexion. It is sufficient to say that the Purulent elements, when thus absorbed or developed, so paralyze the nervous centres and blight the tissues of the organs generally, as to interfere with the action of all the component parts of the organism—suspending nutrition, aborting or altering secretion, robbing the muscles of their tone and power, destroying " nervous influence," and inducing a complete revolution in the whole system.

Organization. The Plastic Lymph effused in connexion with the Inflammatory process may either *Organize* or *break down into Pus*. The term *organization* includes the *conversion* of the *effusion* into *tissue*, taking its character from the subjacent structures; the development of *false membranes*: and the formation of certain *heteromorphous products*, as *Tubercle, Cat* ., &c. Plastic Lymph possesses an inherent capacity for organization. As soon as it is effused, this tendency manifests itself by the formation of cells and nuclei in great numbers, which connect themselves with each other, and gradually spread out into fibres lying for the most part in parallel lines, and profusely inlaid with granules. Vessels soon show themselves, which are the result either of a new epigenesis, or the contributions of the neighboring structures, the latter being the more common source of supply. Nerves and absorbents, finally appear, but whether they are supplied by the surrounding tissues, or are spontaneously developed from the effused matter, has not been determined by Pathologists. In this manner the effused Lymph either assumes the characters and functions of the tissues with which it is in contact, or forms *false membranes*. When, however, there is a deficiency of nervous influence in the part or system, a want of plasticity in the effusion itself, or a deficiency of vital power in contiguous tissues, the same attempt at organization is made, but the issue is an aborption, and a product results, of an inferior degree of organization, and lower in the scale of vital endowment, to which the term *heteromorphus* has been applied. It is in this way that Tubercle and other similar growths are developed, as the effects of Inffammation, complicating the termination of that process, and inducing eventually the most serious consequences to the system.

Inflammation may terminate, leaving behind the higher products thus formed in a state of complete organization, and materially modifying the *structure* and *functions* of the part in which they have been developed. An organ, as the Liver or Spleen, which has been *hypertrophied*, by the organization of Lymph effused into its structure, does not preserve its original *status* either *physically* or *functionally*, and is, hence, modified to an appreciable extent by the precedent morbid action. So likewise, False Membranes, by agglutinating the Intestines, binding together the Costal and Pulmonary Pleura, constricting or contracting the Urethra, &c., &c., materially interfere with the legitimate functions of these parts, and produce disastrous consequences in the economy.

Transformation of Tissues. Each tissue posses the power of appropriating certain elements supplied by the Blood, and of converting them into its own substance. In order that this "*formative power*" may be legitimately exercised, it is necessary that the structures remain in a condition of health, that the ordinary supply of nervous influence and of proper pabulum be supplied them, and that their normal Physiological *status* continue intact. The Inflammatory Process interferes with the supply of nervous influence, and destroys the responsive power of the tissues without *necessarily* depriving them of the elements which constitute their proper food. Instead of converting plastic Lymph into their own substance, they simply impress it with a sufficiency of vital force to insure its organization into tissues of a lower grade, and, hence a species of *degeneration*, or *transformation* ensues—the original elements of the structures concerned being consumed by the destructive Metamorphosis which takes place in them, in common with all the tissues of the organism. The whole process may be thus summed up:

1. Inflammatory action causing a diminution of nervous influence in a given tissue, together with a loss of susceptibility to this influence, and an abatement of its energy.

2. The constant destruction of the original elements of this tissue, in obedience to the general law of metamorphosis which applies to the whole organism.

3. The organization of Plastic Lymph, and its conversion into a tissue of inferior vital endowment.

4. The entire substitution of this inferior tissue for the original one, and the consequent modification of the part both as regards "structure and function."

The most common instances of this *degeneration* or *transformation*, are the cellular, mucous, cutaneous, fibrous, calcareous and fatty.

The Fatty degeneration is the most usual universal and important of all, since there is hardly any organ or tissue of the body in which it may not occur. It has been observed in the Lungs, Placenta, Cartilages, Bones, Cornea, Lens, Arteries, Heart, Kidneys, and Liver—particularly of drunkards—, and constitutes one of the most important products with which the Pathologiest has to deal.

It is important to remember that the various tissues possess different degrees of vitality, some being much more highly organized than others; and, hence requiring dissimalar conditions for the full exercise of that power by which plastic Lymph is converted or transformed into their substance. The products now under consideration, differ from the cacoplastic deposits, referred to in another connexion, in being more highly organized, and in the fact of their requiring the exercise of a greater degree of vital energy on the part of the affected tissues in order to insure their development.

All the facts in regard to the *organization* of the *effusion* incident to the Inflammatory process may be thus arranged:

1. Inflammation—
2. Effusion of Plastic Lymph—

Organizing *completely*, and forming False Membranes, or being converted into the Elements of subjacent tissue.

Organizing *incompletely* and forming tissues of inferior vital endowment to those affected by the inflammatory action.

Organizing *less completely*, and forming Heteromorphous products generally.

Not organizing at all, but breaking down into Pus.

Death of the Part.—Inflammation causes the death of the part, in which it occurs in two ways, not materially differing from each other in their essential nature.

These processes, which are at once the *effects* of the Inflammatory process, and the instrumentalities by which it accomplishes its work of destruction, are Ulceration and Gangrene.

Ulceration.—In atrophy the form and structure of the part remain, while the breaking down and absorption of its elements take place with unusual rapidity—more rapidly, in fact, than they can be reproduced. It sometimes occurs that this power of Inflammation localizes itself, and so completely annihilates the equilibrium between the waste and repair—cell-destruction and cell-elaboration—of a tissue, *within certain prescribed limits*, as to insure the entire *suspension* of that Physiological process by which the structure is *built up*, and to stimulate, to an unusual degree, that destructive metamorphosis by which it is *broken down*. There results, consequently, a possitive disorganization of the part thus affected, with an actual loss of its substance—forming what is familiarly known as an *ulcer*.

It is in this way, that ulcers are originally developed, as the result of Inflammatory action, while the particular features which give them character, are impressed upon them by extraneous circumstances. *Ulceration* is therefore the local, circumscribed destruction of a tissue,—a veritable dissolution in miniature. The several distinct pathological acts concerned in the development of an ulcer may be thus enumerated:

1. An Inflammation which localizes itself.
2. A suspension within circumscribed boundaries of that process by which the tissue repairs itself.
3. An unnatural, morbid, excessive exercise of that process by which the waste—cell-distruction-of the part is accomplished.
4. A consequent solution of continuity, and the presence as effete, extraneous matter of certain

portions of the tissue which have not been so readily or rapidly absorbed.

In support of this view of the subject it is only necessary to mention that the debris of the wasted tissue cannot be found either in the Pus which, subsequently fills the ulcer, or in the Blood itself—a fact which demonstrates that their disappearance is due to an exaggeration of the Physiological process by which the destruction of tissue occurs throughout the whole organism—or in other words that they disappear in obedience to the ordinary law of cell-destruction and absorption enforced with extraordinary energy and effect, through the agency of the Inflammatory Process.

For the different varieties of ulcers, with their symptoms, treatment, &c., the reader is referred to the standard works on Surgery.

Gangrene.—Gangrene may be considered as a partial death—the death of one part of the body while the other parts are alive. It may result from the *violence of the Inflammation;* from *an arrest of the Circulation;* and from *deterioration* of the elements *of the Blood.*

The expression "violence" is used relatively in this connexion—to convey the idea of an Inflammation not intrinsically great, but still too excessive for the part to bear without serious detriment. The same amount of vitality does not reside in all systems, nor is it distributed in equal proportion to the tissues and members of a particular organism. When Inflammation is developed in a tissue or member, the vital force of which has been lowered, and whose "formative-power" is lost, a peculiar modification of structure occurs, to which the term *mortification* is applied. In a word, the tissues not being supplied with their normal amount of vitality because of its consumption by the Inflammatory Process, and having, therefore, lost the power of appropriating the pabulum necessary for their support, die, as a matter of necessity, while the other portions of the body remain intact. Atrophy indicates that the "formative-power" of the tissues has been diminished, while a sufficiency of vitality remains to preserve their external form and internal organization; ulceration shows that this same power has been lost within circumscribed limits, and that molecular death has been the result; while Gangrene illustrates the fact that this ability to "appropriate and transform," has been entirely destroyed, even to the extent of entire tissues and members, with such a diminution of vitality as precludes the preservation of their organization, and permits the operation of ordinary chemical affinities.

The circulation may be arrested by the congestion of a part, and by the pressure of effused Lymph, Serum or Pus. In this way the tissues are deprived of their proper food, and really die of starvation.

The blood may be so altered by Inflammation, particularly of a specific character, as to afford no pabulum to the tissues, and to prove the occasion of their death. Thus the Inflammatory action associated with Small Pox, Scarlatina, Erysipelas, Pyæmia, Hospital-Gangrene, Glanders, and other diseases of a specific nature, terminates not unfrequently in the mortification of some part or member of the human frame.

In some instances the dead portion is dissolved away at its circumference by an exudation from the living parts, and is thus separated or *sloughed* from them; while, if the dead portion be extensive, separation will not be effected before decomposition takes place, and, hence, we have what are known as Gangrene and Sphacelus.

Gangrene may be regarded as the state which precedes and terminates in *Sphacelus*,—a condition in which there is great diminution, but not a total destruction of the powers of life,—the blood still circulating through the larger vessels,—the nerves retaining some portion of their sensibility, and the part being not yet beyond the recuperative point.

By Sphacelus is meant the positive and irrevocable death of the part,—the loss of its organization, the destruction of its component elements, the suspension of its vital laws, and its complete surrender to chemical principles and affinities.

Gangrene has also been divided into the *humid, dry, constitutional* and *local*. But it is not our purpose to consider these varieties in detail, in as much as the same principles are concerned in their development, and similar laws apply to their treatment. When mortification is about to manifest itself as a result of Inflammation, the *redness* assumes a darker hue: *heat* and *pain* abate; and there is a general amelioration of all the symptoms save the *swelling* which generally increases in consequence of the effusion of *Sanguinolent* Serum.

When Gangrene terminates in Sphacelus the hue of part becomes dark and dirty—the tissues grow flaccid and cold, while crepitation manifests itself on pressure, and a most offensive odour is evolved.

When the progress of the Gangrenous process is arrested, healthy circulation is developed up to the margin of the diseased portion, while a bright red line—the *line of demarcation*—indicates the establishment of adhesive Inflammation, and shows that the living parts are to be separated from the dead by a spontaneous effort of nature. This *line of demarcation* extends to the entire depth of the Gangrene, totally and completely surrounds it, and by a process of interstitial ulceration, removes the dead part, without hemorrhage or other serious inconvenience, leaving a granulating and healthy surface behind, which undergoes cicatrization without much difficulty or delay.

In some instances however, as when Patients have been subjected for a protracted period to the influence of debilitating agencies, the blood does not coagulate in the vessels and hemorrhage of a fatal character occurs.

TREATMENT OF INFLAMMATION.—As the phenomena of Inflammation connect themselves both with the *Part* affected and with the *System at large*, it is plain, that the remedies employed in its treatment must be of a *Local* and a *General* character. This constitutes the first and most important classification of the remedial agents at the command of the Surgeon in his contest with this dangerous, and often defiant malady, though more minute subdivisions may be necessitated by an accurate and elaborate investigation of the subject. From the account which has been given, of the symptoms, products and terminations of Inflammation, in the preceding pages of this work, it is plain, that the Remedies employed in its management, should be used with reference to the following Indications:

1. To control the response made by the system at large to the local disturbance—i. e. to control the adventitious,—non-essential phenomena of Inflammation.

2. To control the Heat, Pain, congestion, &c.— the essential phenomena of Inflammation.

3. To limit the effusions incident to the Process—i. e. to confine the Inflammatory Action within Physiological grounds by securing simply the *repair* of tissues.

4. To promote the re-absorption of the effusion and to *restore* the tissues to health—i. e. to insure Resolution.

5. To prevent modifications in the structure and functions of tissues and organs—

6. To prevent the death of the part affected, either molecularly, by *ulceration*, or entirely by *Gangrene*.

All *General* and *Local* Remedies, used in the treatment of Inflammation, act either by *controlling* the phenomena, *limiting* the effects, or *modifying* the terminations of the Inflammatory process.

General Remedies.—Inflammation may be associated with a system in a condition of vigor, or of debility, and is sthenic or asthenic according to the circumstances of the case. When connected with a healthy and vigorous system, it is usually characterized by such symptoms of Inflammatory Fever,—as were referred to under the head of Sthenic Fever, and when developed in connexion with an impoverished and debilitated system, the resulting Febrile action is of a Typhoid character. These facts necessitate a division of the constitutional agents employed in the treatment of Inflammation into Depletory and Stimulant Remedies.

Depletory Remedies.—Among the most prominent agents which belong to this class are Blood-Letting, Mercury, Depressants, Cathartics, Emetics, Diuretics, and Diaphoretics, Nervous Sedatives, Agents which control the Capillary circulation and the Anti-philogistic Regimen.

Blood-Letting.—Without entering into the merits of the great Blood Letting controversy, which has so divided the Medical world, it will be sufficient for present purposes, to mention the circumstances, &c., under which, according to the instructions of the ablest masters, and the teachings of a sound therapy, the Lancet may be employed in the treatment of Inflammations.

1. Blood Letting should never be resorted to save in Inflammation which connects itself with a constitution which is strong and healthy,—that for instance of a vigorous, athletic man.

2. When Inflammation, is associated with Ple-

thora—a full habit, and an unusual supply of red blood.

3. In Inflammations which produce an excessive disturbance in the system at large, accompanied by a full pulse, hot skin, flushed face, and the usual evidences of Inflammatory Fever.

4. In Inflammations of some internal organ, which manifests itself by symptoms of great depression such as small pulse, cool skin, clammy perspiration—in a constitution healthy and vigorous up to the moment of the attack.

5. In all Inflammations of a high grade, when no tendency to Typhoidism exists, and the Patient can be subjected subsequently to proper treatment

The ends which may be accomplished by Bloodletting are:

1. To lessen the amount of blood when it is too great, and to reduce its quality when abnormally rich or stimulant, and thus, to relieve Irritation and Inflammation.

2. To lessen the action, of the Heart and Arteries, to restrain the momentum of the circulating fluid, and consequently to diminish Heat, to abate Pain, to prevent effusion, to equalize the circulation, to obviate local determinations, to relieve spasm and nervous irritation, and to arouse the susceptibility of the various organs, rendered insensible by the congestion of the Nervous Centres.

3. To promote absorption, and to increase the action of other remedies.

4. To arrest Hemorhage by inducing syncope, and favoring the formation clots, by which the

Vessels are blocked up, and the escape of Blood prevented.

It cannot be denied however that there are multitudes of cases, particularly in connexion with the Surgery of Camps, and Hospitals, in which Blood-letting would not be beneficial, but positively injurious. But as a pure anti-phlogistic when the grade of the Inflammation is high and the attendant conditions are such as to admit of its proper application, the Lancet has no rival, particularly if employed before the exudation of Plastic Lymph, or the development of those phenomena which indicate that the acme of the disease has been passed. This can be readily understood, when it is remembered, that in Inflammation, with each pulsation of the Heart an unusual amount of Blood is sent to the affected part, which serves to keep up and to increase the already excited irritation; that the Blood itself is far more stimulant than in health; that the momentum of the circulating fluid, is greatly increased; and that nervous irritation exists far beyond the natural limit,—morbid conditions which Blood-letting ameliorates and removes upon the principles already enunciated.

The employment of the Lancet is based upon the supposition, that, though the nervous centres possess the inherent power of generating a sufficiency of vitality or nervous force, they are prevented from so doing by the presence and pressure of an unusual quantity of depraved Blood, and that the proper performance of their functions can be facilitated by the removal of this pressure, and the supply of a better material for their consumption.

Arterial Sedatives.—Veratrum Veride, Digitalis, Aconite, Tartar-Emetic, &c., produce the same effects as Blood-letting, though in a less marked degree, by the impression which they make upon the circulation. Under their action, the skin relaxes, the pulse softens, the tongue grows moist, secretions are restored, nervous irritation abates, and everything indicates the restoration of the circulation to its normal equilibrium, and the abatement of the Inflammatory symptoms. Their employment is particularly adapted to the cure of Inflammations of an acute character, in young and robust subjects, whose systems require to be rapidly impressed in order to stay the march of the disease. In Inflammations of the Respiratory Organs their beneficial effects are so particularly marked that they have almost entirely superseded the use of the Lancet.

These agents are not *spoliative.* They do not deprive the system of its blood, and thus rob the tissues of their food. Their depressing effects are consequently far more transient than those produced by the Lancet; and, hence there is not the same difficulties to be apprehended in building the system up—in giving it tone and recuperative power—attendant upon their administration, as upon the employment of Blood-letting—a most important circumstance in these times of Typhoid tendencies, and low grades of Fever generally.

Mercury.—This Drug not only *controls* the

symptoms of Inflammation, but *limits* its effects, and *modifies* its terminations.

It *controls* Inflammation by rendering the Blood less irritable; by diminishing the momentum of the circulation; and by promoting the secretions, and by acting as a depletant and a deobstruant.

It *limits* the *effects* of Inflammation by robbing the Blood of its Plasticity, and thus precludes extensive effusions; and, by releaving local congestion, accomplishes the same end.

It *modifies* the *terminations* of Inflammation by *liquifying* the effused lymph and facilitating its absorption—thus promoting resolution, or "termination in health;" by promoting *absorption*, and obviating *induration*; by destroying false membranes,"—thus preventing modifications in the "structure and functions of tissues;" and controlling ulceration by altering the condition of granulating surfaces.

Rules for the administration of Mercury:

1. Administer it in the form of Calomel, Blue Mass, or Mercury with Chalk.

2. When from the violence of the Inflammation, a prompt and powerful impression is required, Calomel should be given in large and frequently repeated doses.

3. When the Disease is less violent, and the organ not important to life, Blue Mass or Mercury with Chalk may be given in smaller doses.

4. To make Mercury more Purgative combine with it finely powdered White Sugar, and give it upon the Tongue.

5. To prevent it from running off the Bowels,

combine with each dose a small quantity of Opium.

6. Never administer Mercury without endeavoring to ascertain if the Patient possess any Idiosyncrasy in regard to it.

7. Do not administer Mercury in any form to persons of a strumous habit, to the very aged or infirm, to those who have been much enervated by the depressing influences of bad clothing, crowded and ill-ventilated tents, and improper food, or to the consumptive.

8. Never produce Salivation designedly, or in other words, suspend the medicine so soon as a free secretion from the Salivary Gl--- shows that the system is saturated with it ate of Potassa, administered in large an... ntly repeated doses, is the best remedy for Salivation.

Cathartics, Diaphoretics and Diuretics are administered for their depleting, derivitive or revulsive effects.

Nervous Sedatives. Although the part played by the Nerves in Inflammation is not thoroughly understood, yet the following facts may be regarded as established:

1. The primary morbid impression is made upon the nerves, from which it is reflected to the Capillary Vessels, and hence *irritation* and subsequently *congestion* are the primary phenomena of the process.

2. The *nervous centres*, responding to the perturbation, thus induced in the economy, participate in the *irritation*, and hence, the circulation and the secretions, together with the nutritive process are disturbed.

3. Inflammation is as much a product or concomitant of nervous irritation, as of vascular disturbance.

It has been shown, that when the Opthalmic branch of the fifth pair is divided in the Cranial Cavity of a Rabbit at the Varolian bridge, Inflammation is developed in the surface of the eye, and that, when the nerve is cut in such a way as to divide the Ganglion of Gasser, the Inflammation is more violent and deeply seated. It has also been demonstrated that when the Pneumogastic Nerves are cut high up in the neck, the Lungs become engorged with Blood and present many of the phenomena of Acute Inflammation, while the stomach becomes also envolved to the extent of an arrest of its secretion. So likewise when the Brachial Plexus is tied, the integuments and finally the deep structures of the Limb become inflamed in a very high degree. These and a multitude of kindred facts which modern Physiology has established, demonstrate that the role performed by the nerves in the development and continuance of the Inflammatory Process is one of the greatest importance.

The Therapeutical action of Sedatives is to diminish the injection of the nervous centres, to relieve the irritability of the whole nervous mass, and thus, indirectly to restrain the action of the Heart, to disgorge the Capillaries, and to regulate the action of the secreting organs.

From this plain statement in regard to the condition of the nerves in Inflammation, and the therapeutical action of Sedatives, it is made apparent that this class of remedies is peculiarly indicated in the treatment of that morbid process.

The agent which stands at the head of this list is *Opium*, with its different preparations, as the Salts of Morphia, Laudanum, and Dover's Powder, though Stramonium, Hyosciamus, Indian Hemp, &c., may also be employed. This remedy is particularly indicated when the Inflammatory Process is accompanied by violent pain, a symptom which may complicate the morbid action to a considerable degree, even to endangering the patient's life. Rules for the administration of Opium.

1. Precede the exhibition of the Opiate, by Bleeding or Purgation, particularly when there is Plethora, Fœcal distention, Disorder of Secretion, &c.,

2. Administer it in large doses—say from two to four grains of Opium within every twelve or twenty-four hours.

3. Give the Opiate at night, so that rest and quiet may be secured to the patient.

4. Remember, that under the influence of Pain the System acquires a greater tolerance for the Opiate.

5. If the skin be dry, combine with the Drug some Diaphoretic or use Dover's Powder.

6. When Inflammation occurs in structures which are likely to be put in motion by the normal processes of the economy, as the Peritoneum, the Pleura, the Alimentary Canal, &c., Opium may be freely used, not only for the purpose of controlling the Inflammation already existing, but to keep the part at rest and thus indirectly to prevent the

farther development of it,—by serving as a veritable splint to the affected structure.

Agents which contract the Capillaries. In those cases where the local disturbance is excessive, accompanied by great Heat, Pain Congestion, and Swelling, it becomes a matter of importance to act upon the Capillaries in such a manner as to limit the amount of Blood in them. The remedies by which this end can be most readily attained are Ergot, Belladonna, and the Muriated Tincture of Iron,—agents which by diminishing the calibre of the Vessels, reduce the local hyperæmia, and relieve the unfavorable symptoms incident to it. Theoretically these Remedies, from their known theraputical properties, would seem to be particularly indicated in the treatment of that variety of Inflammation refered to; but as yet the utility of their administration has not been subjected to that practical test which the Profession demands as the essential condition of its faith and confidence.

It must not be supposed however, that these agents act particularly upon the diseased part, for it is only by the impression made upon the organism in its totality that the engorged Capillaries are incidentally contracted, and the congestion relieved.

They are certainly worthy of a thorough and impartial trial, and as such are recommended to the Profession.

Antiphlogistic Regimen. Under this head are included the diet of the Patient, and certain other circumstances and conditions by which he may be surrounded. During the height of the Inflammation, when the functions are interrupted, the secretions deranged, and the Blood filled with stimulating elements, great care should be observed in the regulation of the diet. Food is usually loathed, under these circumstances, and when injested in solid form, serves only as an additional source of irritation to the system. When the acme of the affection has passed, mild and easily digested food should be administered in a liquid form, beginning with gruel, arrow root, &c., and gradually and cautiously advancing to other more nutritive articles. The drink should be cooling and demulcent. The question of diet in connexion with the Inflammatory Process, is an important and delicate one. Care should be taken to run neither into the extreme of over stimulation nor of too great abstinence; but the Surgeon should remember that there is danger to be apprehended alike from an excessive supply of pabulum to the already infected Blood, and from the debility which necessarily and in some instances, rapidly ensues from the destructive metamorphosis incident to Inflammatory action.

It is also a matter of the first importance to secure perfect tranquility of mind and repose of body, as well as a proper amount of healthful sleep.

So, also, recovery and comfort are both promoted by a regular temperature, free ventilation, cleanliness of body, words of encouragement and kindness, clean and comfortable bedding, the presence of friends and relatives, the assurance of victory, confidence in the skill and humanity of the Surgeon, and a multitude of similar circumstances which will readily suggest themselves to the Physician.

Stimulants.—This class of remedies is indicated:

1. When the morbid action has been originally developed in connexion with a depraved and debilitated system.

2. When the strength of the system has been exhausted by the Inflammation itself or some of its products, as when Hectic is developed in consequence of the excessive discharge of Pus, &c.

The principle upon which stimulants are exhibited, in this connexion, may be thus explained.

The Nervous Centres, which are the great fountains of vitality,—the sources from which flow out the influences which give tone to the muscles, action to the secerning organs, life and power to the whole organism, become debilitated from the absence of those conditions which are essential to the health of the system, such, for instance, as pure and proper food, cleanliness of person, appropriate clothing, contentment of mind, and other similar circumstances. The Blood, at the same time, is impoverished, losing its red globules, augmenting in watery elements, and becoming more irritable and less stimulant to the tissues through which it circulates. The development of Inflammation, finds a system in which these changes have occurred, in but a poor condition to resist its invasion, and to prevent the induction of its most unfavorable consequences. The great centres, from their preternatural irritability, respond immediately and violently to the local impression,— so excessively in fact, as speedily to exhaust themselves, and to lose the power of supplying that influence upon which the integrity of the Organism so much depends. The tissues generally being thus deprived of the stimulus from the Nervous System, are incapable of appropriating their necessary pabulum. The secerning Organs having lost their guiding and controlling principle fail in the performance of their legitimate functions. The chief motive power of the circulation being weakened or destroyed the Heart beats wildly, the arteries contract irregularly, the Capillaries engorge themselves with blood, and the circulating fluid is vitiated to a still greater 'degree under the constantly increasing demands for its vitalizing principles, the retention of the checked Secretions, the development of Inflammatory products, the waste of exhausted tissue, and the expenditure of its Carbo-Hygronous elements in the work of Calorification. In this emergency, the only means of preventing speedy disorganization,—the complete overwhelming of the system by the violence of the disease, is to supply it with strength. The exhausted fountains must be replenished, the wasted stream restored, the motive power of the paralyzed machinery supplied anew, or the Patient surrendered to the embrace of death. Stimulants, therefore become a necessity, so absolute and imperative, that to fail in their employment is to assume the responsibility of a fatal result.

A simple illustration will elucidate the whole subject. The system may be likened to a Fortress, the Inflammation to the attacking Party,—the first resisting the assault, the latter striving to reduce the Work. Now, it is plain that a successful resistance can be ensured in two ways :—either by weakening the attacking Party, or by strengthening the Fortress. By Depletants, we diminish the number and power of the assailants, and thus ensure the safety of the Garrison. By stimulants, we strengthen the weakened Work, victual and encourage its defenders, and secure the repulse of the attacking Party. The one plan prevents fatal consequences by abating the force and intensity of the Inflammation, while the other accomplishes the same end by strengthening the sinking and overpowered system.

A large majority of army Patients present symptoms of debility in connexion with the progress of all Inflammatory affections, and the exhibition of stimulants and tonics is consequently demanded as a general thing in the treatment of their diseases. This is the general rule, but it must be borne in mind that it has its exceptions, and is not of universal application, as some teach and many believe. Many affections which assume a Typhoid type during their progress and thus necessitate a resort to sustaining remedies, may be "cut short" by the timely employment of active measures; while it is possible to stimulate too excessively even in diseases which primarily and unequivocally demand that plan of treatment. It should never be forgotten that debility is as surely and speedily produced by surfeiting the Nervous Centres with too great an abundance of the rich food which alcoholic Preparations supply, and by over taxing their generating properties by excessive and protracted stimulation, as by any other possible means, and that Carbon and Hydrogen, whilst subserving valuable purposes in the economy, are not the elements from which the most important structures of the organism gather their vitality or power.

In exemplification of the truth of these observations, it is only necessary to refer to the recorded experience of Dr. Gualla, Surgeon in chief of the military Hospitals at Brescia, in regard to the treatment of the wounded after the battle of Solferino. He declares that the Italian soldiers who fought upon their own soil, in their native climate, in their full vigor and health, and not exhausted by long marches, or injured by unusual food, recovered rapidly from their wounds, though subjected universally to an Antiphlogistic treatment; while the French soldiers, who were weakened by the dangerous and protracted march over Mount Cenis, suffered greatly and died in large numbers, though treated on opposite principles. The greater suffering and mortality of the latter, he ascribes to the fact that they were allowed too rich a diet, and stimulated to an unreasonable extent, notwithstanding their previous debility. It is much to be apprehended, that under the extravagant teachings of Todd, Bennet, and others of their School, the death blow of many an unfortunate victim has been given in the excessive potations administered for his comfort or relief. The Surgeon should enter upon the discharge of his most responsible task with an honest determination to discard all bias and prejudice in regard to particular modes of treatment, and with "*in medio tutissimus ibis*" as his rule, should only depart from it after a thorough individualization of each particular case, and an accurate knowledge of all the surrounding circumstances.

Stimulants and tonics have been spoken of indiscriminately, for the reason that the one is so administered as to secure *permanency* of impression and the other employed in such a manner as to *ensure rapidity* of action, thus approximating them therapeutically, and making them subserve the same ends in the economy. The Remedies of this class in general use, are Alcoholic Liquors of all kinds, Wines, preparations of Ammonia, Sulphate of Quinia, and Tinctura Muriatici Ferri, &c.

The alcoholic liquors stand at the head of the list, in as much as they can be more conveniently obtained and administered; as their effects upon the system are prompt and decided; as they are better borne by the stomachs of most men; and as they are more palitable and agreeable to a large majority of Patients.

Rules for the administration of alcoholic stimulants—

1. Examine well into the present condition and previous habits of the Patient before administering them.

2. Commence with small quantities—say half an ounce, and gradually increase the dose according to the necessities of the case.

TREATMENT.

3. Administer at such intervals as will ensure a prompt, continuous and equable impression upon the system.

4. Watch the condition of the stomach, carefully, lest an irritation of that organ be developed, thereby interfering with the absorption of the stimulant and adding to the burdens of the laboring system.

5. Examine the Pupil frequently, noticing whether it be contracted or dilated abnormally and discontinuing the Remedy from the fear of cerebral Inflammation in the one instance or a excessive congestion in the other.

6. Attend strictly to the circulation, continuing the medicine, if the Heart beats more slowly under its influence, or continues at its original rate, and rejecting the stimulant when its pulsations are excessively increased in frequency and force.

7. If Coma be produced, Delirium increased, or sleep prevented, change the treatment.

8. If the Tongue grow red and cracked, the mouth dry, deglutition difficult, and the voice husky, stimulants are contra-indicated and should be abandoned.

9. If the Kidneys or Skin—particularly the latter—be too much acted on, thereby debilitating the Patient, stimulate carefully.

10. If the Heat, Pain, and congestion of the part increase, or the wound looks redder, fails to suppurate, or discharges Pus too freely, the remedy should be discontinued.

11. On the other hand, when none of the acci-

dents just mentioned present themselves, and the attendant phenomena assume an opposite character, the Surgeon should not be alarmed at the quantity of the stimulant employed, but being guided alone by its effects, and observing the progress of the case with the most intelligent scrutiny, he should push his advantage until the system has secured an entire mastery of the Disease.

In regard to the particular preparation of Alcohol which should be employed, the fancy of the Patient, or the convenience of the Surgeon may be consulted when *good* Liquors are within reach. Whiskey is usually preferred, because when pure (?) it is more acceptable and less irritating to the stomach; while French Brandy also has its champions. In the present condition of the Country, Apple Brandy is the purest, most palatable, and least difficult to procure, since distillation from grain has been prohibited; while experience has convinced the Author, that as a pure stimulant it stands unequaled.

The manner in which the other Remedies referred to under this head, are employed will be more fully considered in various connexions.

Local Remedies.—These are either *preventive* or *curative*, according to the end for which they are employed. The most prominent and important among them are Rest, Position, Local Depletion, Revulsives, Cold and Warm Applications, Topical Alteratives, and Compression.

Rest.—The importance of steady and persistent rest, can readily be understood, when it is remembered that the least exercise of the part necessitates the flow to it of a large amount of blood and nervous influence. Where Rest cannot be secured by the Patient's own efforts, Splints may be employed, or Opium used for the purpose of temporarily paralyzing the muscular fibres of the affected structures. It is well not to continue this treatment long, lest anchylosis, permanent immobility, &c., be the consequence.

Position.—In Inflammation the vessels are filled with an unusual amount of Blood, which is still controlled by the laws of gravity,—accumulating in a dependent part, and *vice versa*. This is true for the other fluids as Serum, Lymph and Pus which are developed in connexion with the process of the morbid action. So likewise position may increase the pain of the affected part, by causing muscular pressure upon it. For these reasons, the part should be kept in an elevated position, and so arranged as to relax its muscles, while the comfort of the Patient should likewise be consulted as far as practicable.

Local Depletion.—This is accomplished by means of Scarifications, Punctures, Leeches, Cups, and Drainage. The Blood may be taken directly from the part by local bleedings, or robbed of its serum by Blisters. These remedies also exercise an indirect control through the agency of nervous reflex action, or by their general sedative effects upon the system at large. Blisters should always be employed with caution particularly in the earlier stages of Inflammation, lest they add to the irritation of the diseased structure, and thus prove an injury rather than a benefit to the Patient. Punctures are employed for the purpose of relieving the suffering tissues of the Serum or Pus which may have been poured out in them. The artificial evacuation of Pus may be accomplished either directly by the Knife, by Caustic—though the latter is seldom attempted—or, by what is known as Drainage.

Rules for opening Abscesses.—1. Take care in introducing the Bistoury not to interfere with any important nerve, to open a large vessel, or to penetrate one of the large cavities of the Body.

2. Make the opening great enough to ensure a free vent of the pent up fluid. It is far better that the opening should be too large than too small.

3. Assist the evacuation if necessary, by the hand or finger, used however, with the greatest gentleness.

4. Prevent the incision from healing by "First Intention," by inserting a small tent made of old linen, well oiled, and interposed between the edges of the wound.

5. Employ the "warm water dressing" or an Emollient Cataplasm(?) for the purpose of promoting the discharge of the fluid, after the bleeding has ceased, but be careful not to continue it for too long a period lest too much relaxation ensue.

6. Approximate the sides of large abscesses by means of compresses.

7. Make the opening, if possible, so that gravitation will promote the escape of the purulent matter, but if this cannot be effected, try a counter-opening. A current of water passed from one opening to the other is frequently of great advantage.

8. If arteries be divided in the operation, ligate them.

Chaissaignac has proposed to relieve abscesses of their contents by a system of "Drainage," which is, in fact, but a revival of the old doctrine of the Seton. He plunges a Trocar lined with a Canula through the abscess and out again through the integument; then withdrawing the Trocar, he passes through the Canula, which remains behind, a tube of India Rubber, perforated with holes for the escape of the matter, and ties the two ends together. In this way the escape of Pus and Serum is facilitated, and a collapse of the parts secured, while the introduction of atmospheric air—an agent which promotes suppuration while it decomposes Pus—is entirely prevented.

Cold and Warm Applications.—This class embraces everything from the "Cold water dressing" to the "Medicated Poultice." Though Cold Water has been used in the treatment of Inflammation from the earliest times, the experience of Ambrose Paré has mainly contributed to the elevation of the remedy to its proper position in the estimation of Military Surgeons. It has perhaps been more universally employed, in the War which is now being waged by the Confederacy than in any other previous struggle, and with results, which when properly tabulated, will astonish the world. To Surgeon J. J. Chisolm, Professor of Surgery in the Medical College of South-Carolina, and Author of the best "Manual of Military Surgery" that has been published in any language, we are indebted for the just appreciation in which this invaluable mode of treatment is held, at the present time. If he had done nothing more than inculcate, in that able work, his most scientific and rational views in regard to the Cold Water treatment of wounds, he would deserve the lasting gratitude of the Profession and of the Public.

Truly it may be said, the days of Cerates, Ointments, and Cataplasm has passed,—having been swept to oblivion by the copious streams of Cold Water, with which an enlightened Surgery has comforted and relieved the mutilated victims of a thousand Battle Fields.

The advantages of "Cold Water Dressings" in all stages of Inflammation, as local applications, may be thus summed up:

1. Cold Water is clean, cheap, simple and generally agreeable to the feelings of the Patient.

2. It enables the Patient himself, or the most ignorant assistant, to dress the wound.

3. It keeps down the temperature of the parts, constrains the Capillaries, and relieves Hyperæmia.

4. By forcing the Blood out of the Capillaries and preventing its passage into them, the source from which Pus is developed, is thus cut off, and the suppurative process arrested. It has been conclusively demonstrated that the process of suppuration, so far from being necessary to the healing of wounds, or the arrest of Inflammation, retards the one and seriously complicates the other. The importance, therefore, of Cold Water Dressings, even in the most advanced stages of Inflammation is thus made apparent.

6. It relieves nervous irritation, and thus, both directly and indirectly controls the Inflammatory process.

Cold Water may be applied in various ways, as by saturating Linen, Cotton Cloths, Sponge, &c., and frequently squeezing them over or constantly applying them to the part; by suspending a Bucket and then by means of a narrow strip of Cloth, on a lamp wick, conducting a stream of Cold Water to it; by elevating a funnel above the part affected, filling its nozzle with lint and permitting the Water to percolate through it from above; by means of Bladders filled with pounded Ice; and by many other contrivances which the circumstances of the case will suggest to the Surgeon. Care should always be taken to prevent the bed clothing, and the clothes of the Patient from becoming saturated, lest he be chilled or inconvenienced thereby. In some instances, though they are rare, Cold Water cannot be borne at first, when Tepid Water should be substituted for it temporarily, taking pains, however, to lower the temperature of the application, gradually but decidedly, until that degree has been reached at which "sedation and astringency" manifest themselves.

Warm Applications.—The circumstances under which Warm applications are demanded, may be thus stated :

1. When the Blood has so completely stagnated at certain points, as to become insensible to the *vis a tergo* warm applications may be sometimes employed, for the purpose of adding to the volume and force of the Blood current flowing towards

the part, and of thus indirectly relieving the Capillary congestion.

2. When an unusual amount of irritability exists in the nerves of the affected tissues, manifesting itself in great pain, tenderness, sensations of cold, spasm, &c., warm applications are indicated, in as much as the resulting Hyperæmia though inevitable is the lesser of the two evils.

3. When, from the extreme delicacy of the Patient's organization, his tendency to pulmonary irritation, the existence of bronchial affections, or the impossibility of making cold applications with that regularity and system requisite for the preservation at an equable temperature, the "Cold Water" treatment is countra-indicated.

4. When the part affected assumes a glazed, red, dry and angry appearance, manifesting no disposition to heal by the "First Intention," resisting the application of Cold Water, and progressing but slowly towards any termination, warm applications may be employed with advantage.

5. When a wound which has suppurated freely suddenly ceases to do so, without indicating a tendency to heal, either with a total abolition of the sensibility of the part, or an extraordinary augmentation of it, Warm Water may take the place of Cold.

The instances, however, in which Warm water is required to the entire exclusion of Cold application, are of comparatively rare occurrence; and the Surgeon should hesitate and most carefully consider the indications before concluding to make the substitution. If there be any doubt in regard to the matter, give the Patient the benefit of it, and continue the Cold Water.

It should not be forgotten, that the exposure of wounds in which Inflammatory action has been developed, to vicissitudes of temperature is the prolific source of Tetanus; and hence, whether Cold or Warm Water be selected, use it freely and persistently—in such a manner as will maintain an equable temperature in the part.

Should it become necessary to employ a poultice, a soft wet Compress, covered with oil silk, and secured by a flannel roller or outer Compress, will fulfill any possible indication. The proper thermal status is preserved by the absorption of animal heat, the application is light and comfortable, medication can be readily effected, the materials are always attainable, cleanliness can be invariably insured, and, in fact, so many advantages present themselves in connexion with it as to "preclude all substitutes."

Either cold or warm water can be medicated, with Sugar of Lead, Sulphate of Zinc, Tannin, Spirits of Camphor, Preparations of Opium, Tincture of Arnica, and according to the presenting indications. The temperature of the Water can be lowered by the addition of Alcohol, common Salt, or a strong solution of Hyperchlorate of Ammonia and Nitrate of Potassa.

Revulsives.—*Ubi irritatio ibi affluxus est* is a pathological axiom, and upon it the whole problem of the revellent action of Medicines is based. The system possesses but a definite amount of Blood and

Nervous force, and by securing their accumulation at one point, all other parts are relieved of them to a certain extent. Counter irritants are employed in the treatment of Inflammation for the purpose of creating a new disease, which, by attracting the Blood, &c., to itself may serve as a diverticulum for the part originally affected. Great judgment is required in determining *where* and *when* to apply them, since if not wisely employed they increase rather than abate the morbid action. As a general rule they should not be used until after some preliminary depletion has been practised, while care should be taken not to apply them too near a delicately organized structure, or too far from one of a different character. To this class belong Rubefaciants, Blisters, and Suppurants.

Local Alteratives.—The most prominent of these are Nitrate of Silver, and Iodine. The first named is used extensively in acute Inflammations as a topical antiphlogistic agent, while the other is more particularly employed to promote the reabsorption of Lymph, as in dissipating a gathering abcess, and softening an indurated tissue.

Nitrate of Silver is not only a powerful vesicant or destructive, but by substituting a new and more controllable action of its own for the one existing in the part, it serves as a valuable auxiliary in the treatment of Inflammation. Its action may be thus stated—

1. As a vesicant, producing counter irritation, and controlling Inflammatory action, as explained above.

TREATMENT.

2. Destroying tissue and thus assisting nature in her work of elimination, as in Gangrene.

3. Neutralizing certain causes of Inflammation, as virus of the serpent, the poison of the cadaver, and thus indirectly restraining that morbid process.

4. Producing certain changes in animal structures, such as prevent the progress of Inflammation; and hence used in Erysipelas and Hospital Gangrene.

5. Substituting a new and more manageable action of its own for the existing Inflammation,— as in Gonorrhœa, &c.

Iodine acts upon the absorbent vessels, and stimulates them to a more vigorous discharge of their duties; and like all other alteratives, controls the great work of cell-development, and cell-destructive,—the metamorphosis, and nutrition of the tissues.

Compression.—The afflux of Blood to an inflamed part may be prevented by mechanical means, and that local congestion prevented from which effusion ensues with its attendant consequences. Nor is this all, spasm may be controlled in this way, the absorbent vessels stimulated, the affected structures supported, and effusions prevented,—results of the most vital importance to the individual parts and to the whole organism. The means of Compression are the common bandage and adhesive plaster, so applied as to make gentle and agreeable pressure over the whole of the affected structures.

In all that has been said as regards the phenomena, pathology and treatment of Inflammation, a direct reference has been had to the acute form of that affection, as it is with that variety particularly that the military Surgeon has to contend alike in Camp, Field and Hospital.

CHAPTER II.

AMPUTATIONS IN GENERAL.

VARIETIES OF AMPUTATION.—Amputations are *Primary* or *Secondary*, according to the period at which they are performed.

Primary Amputations are those undertaken for direct injury, and are performed either immediately after the wound has been received, or after recovery from the *Shock* and before the development of Inflammation.

By the term *Shock* is meant that condition of the nervous system which sooner or later ensues upon particular injuries in certain persons. It is characterized by coldness of the surface, pallor, tremors, an anxious expression of the countenance, small, irregular and feeble pulse, sighing respiration, partial or complete paralysis of the bladder, mental disturbance and incoherence of speech. This condition may continue for a longer or shorter period, but usually disappears in a few hours; while the intensity of the shock is not always in direct proportion to the extent or severity of the wound, as it is sometimes very great even where the injury is trivial. If it persist however, whether the injury be seemingly great or small, there is always danger to be apprehended, and the Surgeon should prepare himself to meet it.

The evidence of the English Naval Surgeons, as summed up by Hutchison, when taken in connexion with that supplied by Macleod from his Crimean experience, clearly establishes the fact that the condition which is known as Shock is not necessarily established immediately upon the receipt of the injury, but that an interval ensues which differs in duration according to the severity of the wound, the agency producing the injury, or the constitutional status of the sufferer.

The Circumstances under which immediate Amputations are demanded are :

1. When the *Shock* is delayed. The importance of seizing upon the moments of comparative tranquility which frequently elapse between the receipt of the injury and the development of *Shock*, was first recognized by Ambrose Pare and Richard Wiseman, and is now growing into favor with the Profession.

2. When the Nervous Depression is slight, or is not developed at all, as sometimes occurs.— Larrey declared that he had lost a great number of Soldiers by delaying the operation too long, within the first twenty-four hours, and recommended Amputation as one of the surest means of relieving the " commotion," and of diminishing its dangers. When the Shock is slight it certainly should constitute no contra-indication to the use of the knife.

3. When a limb is either nearly or completely

torn off, and a dangerous hemorrhage is occurring which cannot be arrested.

4. When the smaller members, as the fingers or toes are seriously injured.

5. When Broken Bones, Fragments of Shell, splinters, clothing, or other foreign substances are lying in the track of the wound in such a manner as to preclude their extraction, and to induce such an amount of pain and nervous commotion generally as threatens the immediate destruction of life. In cases like these, even the teachings of Larrey may be followed and the operation performed whether the shock exist or not.

In deciding upon the practicability of this operation, it is important to take into consideration the *moral* condition of the Patient. All Army Surgeons know that men are brought from the field of battle, either enthused by the combat, indifferent to all save the "fate of the day," and willing to submit to any thing which gives a promise of future revenge, and the prospect of a participation in the triumph of their comrades, or dispirited, disheartened, and depressed, both physically and mentally by the appearance or the pains of their wounds and the idea of permanent mutilation. The meanest coward, under the strange infatuation of the Battle Field,—the roar of Cannon, the flashing of the deadly Bayonet, the deeds of daring done round him, the stirring notes of command, the presence of his comrades and all the wild excitement of the stirring scene may forget his own mortality and be transformed into a hero, who, while the fit is on him, will despise the steel of the Surgeon as thoroughly as the Bullets of the foe. On the other hand, the Soldier who has marched to the cannon's mouth, insensible to fear, and dreaming only of revenge or triumph, is paralyzed by the flowing of his own blood, and is borne to the rear in mortal terror of an operation by which additional pain is to be inflicted or deformity entailed upon him. In the one case, the Surgeon could perform the operation with impunity, while in the other still greater depression would follow each stroke of the knife, and perhaps speedily terminate his existence.

Upon general principles, it might be supposed that the soldier would bear an amputation better when "heated and in mettle"—when excited by the combat and within sound of the cannon, provided he be not completely prostrated by the shock—than when opportunity had been offered for the abatement of his excitement, and calm reflection upon the dangers and inconvenience of the loss he is to sustain; and hence, if it be not contra-indicated by other circumstances, an *immediate* operation may be resorted to.

The circumstances which demand the performance of amputations—*subsequent* to the abatement of the shock and *prior* to the development of Inflammation, are:

1. The tearing off or crushing of an entire limb, without the accompaniment of an uncontrollable Hemorrhage, and with the complication of nervous shock.

2. Compound or multiple fractures, especially of the lower extremities, accompanied by great

laceration of the soft parts, such as amounts to their pulpification.

5. Complicated Fractures, involving the section both of the chief Vessel and Nerve of the member.

6. Simple Fracture complicated by the opening of one of the large articulations, and the tearing of its ligaments.

7. Great injury of soft parts unaccompanied by fracture, with the division of their main arterial trunks or nervous filaments.

8. Extensive destruction of the integuments, such as precludes the possibility of cicatrization within a reasonable time.

9. Fractures accompanied with extensive contusion, generally demand Amputation. Extensive contusion necessitates amputation more than open laceration, even of great extent.

A complete revolution has taken place within the present century, in regard to the advantages of Primary when compared with Secondary operations. The opinions of Faure have been entirely overturned by the more philosophical views of Boucher, dispite the decision of the French Academy; and the most enlightened experience has indicated the wisdom and humanity of resorting to early Amputations, especially in military surgery.

When the circumstances of the case do not justify an immediate operation, the Surgeon should administer a cup of cool water, then wine, brandy or food if possible, and dress the wounds temporarily, waiting for the establishment of reaction before proceeding to take the proper steps for the performance of the later Amputation.— Words of encouragement and kindness, whether the sufferer be friend or foe, should never be neglected by the medical officer, as the *moral* condition of the patient plays a most important part in relieving the nervous depression incident to his physical mutilation.

Reaction should occur within 48 hours after the receipt of the injury, even in the worst cases, and Primary Amputations are supposed to be performed within that time.

Rules for performing Primary Amputations.—
1. Operate within forty-eight hours after the receipt of the injury.

2. Operate as far from the trunk as possible, as every inch saved diminishes the risk of the patient's life.

3. Operate as soon *after* recovery from Nervous Shock, and as much *before* the development of Inflamatory reaction as possible.

4. Operate at a joint rather than go beyond it.

5. Keep the patient under the influence of Chloroform no longer than is absolutely necessary.

6. Cut rapidly, tie quickly, dress slowly, and bandage lightly.

7. Guard against the development of nervous depression, or of excessive vascular reaction, and stimulate or deplete according to the necessities of the case.

8. Let the knife follow the condemnation of the limb as speedily as practicable.

9. Operate upon the lower limbs for injuries

which would not demand the condemnation of the superior extremities.

There are various other principles which should guide the Surgeon in the performance of these operations, but as they apply with equal force to all Amputations, they will be considered under a different head.

SECONDARY AMPUTATIONS.—Secondary Amputations as distinguished from Primary, are those performed after the Inflammation which supervenes upon the injury, has been developed.

There are two varieties of Secondary Amputation, viz:—those which are performed *before* the Inflammatory action has abated—during the Inflammatory Fever: and those which are performed *after* the subsidence, partial or complete, of the Inflammatory action, and in connexion with some of its products, particularly Pus.

The Circumstances which justify Amputations during the existence of Inflammatory Fever, are:

1. Excessive and uncontrollable Hemorrhage occurring at that period.

2. Tetanic symptoms manifesting themselves in connexion with the wound, and resisting all remedies.

3. Indications of a tendency to debility rapidly and unexpectedly showing themselves.

4. A sudden necessity demanding the immediate removal of the patient, when it is manifest that the dangers incident to transportation are greater than those of the operation.

The impropriety of operating as a general rule, before the subsidence of the Inflammatory process—when a great amount of perturbation exists both in the Nervous and Vascular systems—is too plain to require demonstration. Cases, however, do present themselves when the risk must be taken, and the Amputation performed without regard to the principles which ordinarily control the Surgeon in this connexion.

It is impossible to establish any definite and universal law in this regard; and the Surgeon can only be enjoined to individualize each particular case, weighing the danger of delay, thoroughly comprehending the risk of the operation, and duly estimating the nature, extent and potency of the emegrency before him.

These operations sometimes do astonishingly well, surprising both operator and patient by progressing speedily and surely to a favorable termination.

The Circumstances which necessitate Secondary operations proper, or those undertaken *after* the abatement of the Inflammatory process, and the formation of its products, are:

1. Secondary Hemorrhage, occurring at a late period in the history of the case.
2. Rapid and excessive formation of Pus, jeopardizing the life of the patient.
3. Mortification rapidly developing itself.
4. Rapidly decreasing strength of the patient.
5. Necrosis and malignant diseases of bone, or extensive and exhausting ulceration of the Soft Parts, defying other remedies.

SECONDARY AMPUTATIONS.

6. Diseases of the Joints, especially those of a malignant character.

7. The appearance of Tetanic symptoms.

These contingencies and various others of a similar character, which will readily suggest themselves to the mind of the Surgeon, will justify him in resorting to Secondary operations.

Rules for the performance of Secondary Amputations.—1. Operate as far from (above) the seat of injury as practicable.

2. Operate *above* a joint rather than at it.

3. Operate before the strength of the patient is too much exhausted, and before Pus has been absorbed, if practicable.

4. Operate, as a general rule, after the subsidence of Inflammatory Fever, the development of a free and healthy suppuration, and the restoration of the skin to its normal functions, particularly in the affected limb.

5. Operate in cases of traumatic gangrene—save in frost-bites and burns—just so soon as the first symptoms show themselves: but in constitutional or idiopathic gangrene, wait for the line of demarcation.

6. Avoid hemorrhage, during the amputation, as far as possible, support the strength of the patient and watch carefully for symptoms of purulent absorption.

GENERAL OBSERVATIONS.—The great question of the comparative value of Primary and Secondary Amputations in Military Surgery, may be regarded as definitely settled. The experience of the

English, French, and Russian Surgeons in the Crimean war, establishes beyond the possibility of refutation that the results of Primary are far more favorable than those of Secondary Amputations. Thus Macleod declares, "the experience of the Crimea in favor of early operation was unequivocal in both armies." M. Salleron, who was in charge of the Dolma Bagtche Hospital at Constantinople, asserts that from the 1st of May to the 1st of November, 1855, "the total number of Amputations was 639, ie. 490 Amputations in continuity and 149 disarticulations. Of the 639, 419 were Primary operations, furnishing 221 recoveries and 198 deaths; 220 were Secondary, furnishing 73 recoveries and 147 deaths. Thus among the 639 cases there were 294 recoveries and 345 deaths, the Primary operations yeilding more than *half* of the cures, and the Secondary operations yielding not a *third.*" The Russian Surgeons report that "they lost *two-thirds* of all their Secondary operations, but saved *a fair* number of their Primary." So likewise after the battle of Solferino, according to Dr. Gualla, Surgeon in chief of the Austrian Military Hospitals in Brescia, Amputations performed shortly after the injury was done, showed a much more favorable issue than those undertaken later in the stage of advanced suppuration, the proportional result in favor of life was nearly two to one." Chisolm asserts that the experience of every battle field shows that the mortality following the Amputation of limbs, which require immediate operation is *always* less than those performed some days after the infliction of the wound.

From the statistics on file in the Surgeon General's Office, it appears that there were performed in and around Richmond, from June 1st, 1862, to August 1st, 1862, 272 Primary Amputations—furnishing 190 recoveries and 82 deaths, and 308 Secondary amputations, giving 145 recoveries, and 163 deaths. For fuller information in regard to this subject the reader is referred to the tables which constitute the Appendix to this work.— These tables were prepared with great care by Surgeon F. Sorrell, Inspector of Hospitals for the City of Richmond, Va., under the immediate supervision of Surgeon General S. P. Moore, C. S. A., and constitute an invaluable contribution to Surgical science. In thus collecting and perpetuating these important facts, they have stamped their names in indelible characters upon the professional history of the times.

The term "Intermediate" is employed in these tables evidently to designate operations which were performed neither immediately subsequent to the shock, nor after the development of suppuration; and it consequently corresponds with what has been treated of in these pages as a Secondary Amputation made in the *first period*—during the existence of Inflammmatory reaction.

It is true that these views in regard to Primary Amputations, are opposed by many Surgeons of great ability, such as Faure, Hunter, Percy, Blandin and Mause, but they are mainly of the last century, and the weight of authority preponderates greatly upon the side of Primary Amputations in military surgery.

Dr. John Stone, after a most careful and elaborate investigation of this subject, both in its military and civil practice, thus sums up his conclusions:

1. Primary Amputations of the upper extremities are more successful, and to be preferred both in military and civil surgery.

2. That in military surgery, Primary Amputations of the lower extremeties are twice as successful as Secondary.

3. That in civil Surgery it is immaterial whether Primary or Secondary Amputations of the lower extremeties are resorted to.

4. That Secondary Amputations of the upper extremities in civil surgery are 8 per cent. less fatal than in military surgery.

5. That Secondary Amputation in civil surgery are 12 per cent. less fatal than in military surgery.

There is another view of this subject which should not be overlooked. The number who survive after a given number of Primary and Secondary Amputations does not afford a proper index of the relative value of the two operations. The most severely injured have their limbs removed early; while the milder cases are reserved for Secondary Amputation. In estimating, therefore, the value of the two operations, an account must be taken of the more unfavorable circumstances which practically surround early operations, rendering death inevitable in many cases, or materially retarding recovery in others.

So likewise the question is not whether a hundred men freshly wounded and requiring Amputation are more likely to survive than a hundred who have gone through the dangerous ordeal of a Hospital; but whether the first hundred would live to that period—the probability being that they would not.

In view of all the facts of the case, the conclusion is inevitable, that Primary Amputations are far more successful than Secondary, and that humanity and science unite in demanding their performance whenever practicable.

Instruments.—If possible the Surgeon should have always on hand and ready for use, Knives, Saws, Forceps, Tenaculæ, Bone Nippers, Sponges, a Retractor, Threaded Needles, Adhesive or Isinglass, Plaster, a Tourniquet, Cold Water, Brandy, and Chloroform. Amputations, however, have been performed with no other instruments than a well-sharpend Carving or Bowie Knife, a common Saw, and a Fork bent in such a manner as to serve as a Tenaculum; and no Surgeon will ever permit his patient to die from an exhausting and uncontrollable Hemorrhage or any simmilar accident, for the want of an operation, when these implements can be procured.

It has become fashionable to decry the Tourniquet, and many repudiate it altogether. The abuse of a thing is no argument against it when properly used; and, though there are many and serious objections to this Instrument as ordinarily employed it may be made to subserve most important purposes in the hands of a judicious operator. It is true that it only controls the bleeding when tightly applied, and that when so applied it acts as a general ligature around the member, and can be used but for a short time without injury to the limb, and, yet, within the brief period in which it can be used, the fate of the patient may be decided. The experience of every practical Surgeon will confirm the assertion, that, in multitudes of instances, either from the ignorance, fright or fatigue of the assistant engaged in controlling the Artery, or from some sudden spasmodic motion of the patient himself, the Vessel slips from beneath the compressing finger and permits the escape of that precious fluid, whose every drop is required by the necessities of the weakened system. To find a new assistant may be difficult, to search again for the Artery in the midst of the patient's struggles requires time, while but a few turns of the Screw, if the Tourniquet has been previously adjusted, will obviate all occasion for delay, and by arresting the flow, snatch the patient from the hands of death. Again, it frequently happens, that in consequence of some abnormal development of the Vascular System either congenital or the product of morbid action, sufficient digital compression cannot be made to prevent the flow of Blood, and the Tourniquet comes in as a most valuable auxiliary in the arrest of the inevitable and most destructive Hemorrhage which follows the Knife.

It is therefore best not to discard the Tourniquet

entirely but to adjust it upon the Limb *loosely* and yet in such a manner as to enable the Operator to command the artery in an emergency, so that in the event of the accident referred to, the screw may be turned and the Hemorrhage arrested.— This plan is entirely practicable, and, while it obviates the objection to the instrument, fulfills a most important and preponderating indication.

Chloroform.—The discovery of the anæsthetic effects of Chloroform is the great surgical achievement of the age. Under its soothing influence operations have been performed which otherwise would have been impossible, while the amount of suffering obviated cannot be estimated in words. It has thus extended the domain of Surgery, crowned the noble science with fresher and prouder laurels, and proved a source of incalculable comfort and security to the human race. That accidents of a serious character have attended its administration, and that it does occasionally produce fatal consequences cannot be doubted. And yet, when the immense advantages which it secures, both to the operator and the patient, together with the comparative in-frequency of these unfortunate results, are taken into the account, the arguments so speciously urged against its employment, are rendered utterly nugatory and abortive.

Thus Velpeau declares that he has employed Chloroform more than five thousand times without a single accident. Baudens affirms that the French Surgeons in the Russian war, "had no fatal accident to deplore from its use, although it was employed thirty thousand times or more."—

Macleod states that though almost universally employed by the English Surgeons in the Crimean Campaign, but one fatal result could, with any show of fairness be attributed to it. At Guy's Hospital, Chloroform was used 12,000 times before there was any serious accident. Dr. Gross says that he has given Chloroform for more than ten years without an unfavorable result in any case.

In a word, it has been demonstrated by the stern logic of facts that this invaluable agent is far less dangerous than was supposed in the earlier days of its history—less dangerous than many other remedies which are daily used without stint or limitation by those who most bitterly and pertinaciously oppose the administration of Chloroform.

The advantages which attend its administration are:

1. The abolition of all pain—a fact which improves the *moral* condition both of Operator and Patient, with reference to the operation.

2. The induction of a condition of tranquility, in which the muscles are passive, all motion suspended, and the patient is entirely under the control of the Surgeon, so that more difficult, protracted, and nicer operations can be performed.

3. The suspension of sensibility permitting the more thorough examination of wounds.

4. The arrest of Hemorrhage, during the operation. M. Chassaignac has particularly called the attention of Surgeons to this important fact. According to the observations of this distinguished operator, both the Arterial and Venous circulations are materially controlled by this agent—a

conclusion which must be sustained by the experience of medical men generally whose opportunities for an investigation of this subject have been sufficiently extensive. In order to give a rational account of the forces, in virtue of which these phenomena take place, it is only sufficient to compare the condition of a patient operated on in the ordinary state, with that of one under the anæsthetic influence of Chloroform. In the one the apprehension of the operation about to take place increases the number of pulsations, and augments the cardiac impulse; while the disturbance of the respiration, and the efforts made by the patient, when restrained by the Assistants, retards the free return of Venous Blood. An increase in the force and frequency of the pulsations of the Heart, and a retardation of the Venous flow are the circulatory conditions of those who are submitted to operations without the employment of anæsthetics. When Chloroform is administered there ensues a diminution in the frequency and force of the pulse, together with an establishment of the normal condition of the respiration, and the induction of a state of perfect tranquility and quietude.

This fact should be remembered by the Surgeon in connexion with the application of dressings, as the chances of ulterior hemorrhage are greater in proportion as less Blood has been lost during the operation.

Rules for the Administration of Chloroform.—
1. Place the Patient in a recumbent position, so as to maintain a proper circulation in the Brain, and

as a means of avoiding the disadvantages of great muscular relaxation.

2. Remove all causes which are likely to interfere with the Respiratory Function by controlling the Diaphragm—such as tight clothing, heavy covering, sword belts, &c. Upon the same principle, avoid the administration of Chloroform on a full stomach.

3. See to the introduction of an abundant supply of Atmospheric Air with the vapor of the anæsthetic. This is absolutely indispensable to the safety of the Patient, both as a means of preventing too great an accumulation of Carbonic Acid in the Blood, and the possible development of Carbonic Oxide. All the Inhalers which have been invented are objectionable on account of their inconvenience and the difficulty of obtaining a proper supply of Atmospheric Air. The best mode of administering Chloroform is by means of a cloth, folded in the form of a cone, in the apex of which a small piece of sponge is placed. This, impregnated with a drachm of Chloroform, should be held over the mouth and nose, at a distance of about two inches, being gradually approximated until within one inch of the face, beyond which it should not be carried. Great care should be taken not to force the cloth down upon the face of the patient, lest respiration be interfered with and suffocation ensue.

4. In Primary Amputations particularly, and in those undertaken after the development of Pus, precede the anæsthetic by an ounce of brandy or

whiskey, and repeat the dose if the pulse becomes very weak.

5. See that complete anæsthesia be induced and kept up until the operation is completed, but watch the pulse and breathing carefully lest it be carried too far. This should be made the special business of the assistant to whom the Inhalation is confided, taking care to select an experienced man for this purpose.

6. Discontinue the remedy temporarily when the breathing becomes noisy, when the insensibility of the skin is lost, and when muscular power is abolished—as is shown by the falling of the arm or leg when raised from the bed, the dropping of the eyelid when opened, &c. If, in connexion with these phenomena, the pulse does not become too feeble, anæsthesia is perfect and the operation may be performed without the fear of fatal consequences.

Much discussion has taken place in regard to the quantity of Chloroform which should be administered; but, as the strength of the article materially varies, and as the susceptibilities of patients differ widely, it is plain that no specific quantity can be fixed upon in this connexion. Let it be freely but cautiously administered without regard to the quantity consumed, and with an eye single to the effects produced. When much blood has been lost, absorption is more rapid, and a smaller quantity is required.

A consideration of the *modus operandi* of Chloroform does not legitimately pertain to a work of this description; and it is therefore only necessary to remark that its ultimate effects are those of a powerful sedative to the Nervous System—a fact which should never be forgotten or disregarded.

Notwithstanding that sedation is induced by it in the system and the dangers which attend its administration on that account, its usefulness is particularly apparent in connexion with operations performed immediately upon the receipt of injuries. Under these circumstances, it seems to ward off that commotion of the nervous system which we denominate *Shock*, and to prevent the fatal consequences incident to that condition, by taking possession of the great centres and appropriating them exclusively to its own purposes.

Disastrous Effects.—Though Chloroform has proved an inestimable boon to the human race, it is potent for evil also. Effects sometimes follow its administration which all the skill of the Surgeon cannot restrain, and which necessitate the speedy sacrifice of the patient's life. The modes in which these unfortunate results are produced are:

1. By an interference with the functions of the brain—either from the congestion of that organ or the presence in it of too much impure blood.

2. By an interference with the respiratory function. This results from the sedative impression made upon the nervous centres which preside over that important function, and the interruption of the process of pulmonary *osmosis* upon which its integrity depends.

3. By inducing certain alterations in the blood. The changes in the blood which attend the inhalation of Chloroform are the accumulation in it of

an unusual amount of Carbonic Acid, the absence of a proper amount of Oxygen, and the development of Carbonic Oxide. The accumulation of Carbonic Acid, and the absence of Oxygen, which is the vitalizing element of the tissues, can be readily understood when it is remembered to what an extent the respiratory function is interfered with; while the development of Carbonic Oxide can likewise be readily explained. The tissues, through which this altered blood circulates, have a natural affinity for Oxygen—an affinity which increases in intensity just in proportion as there is a deficiency of it. So imperative does their demand become, that the Carbonic Acid gas parts with one of its elements of Oxygen, and is thus converted into Carbonic Oxide—a most deleterious compound according to universal experience. Bernard has shown that Carbonic Oxide has a greater affinity for the blood corpuscles than Oxygen itself, and that it forces them to surrender all of this vitalizing principle, only to become inert and effete matter themselves, incapable alike of stimulating the centres, of supplying the tissues, and of performing their appropriate part in the economy.

4. By interfering with the action of the heart, Dalton has shown by a series of interesting and conclusive experiments that Chloroform kills in a majority of instances by an instantaneous and direct paralysis of the heart—a conclusion which has been verified by other able Physiologists. This demonstrates the importance of watching the pulse as well as the respiration, and of carefully noting all the changes which take place in its

rhythm, rate and volume, during the performance of an operation.

From the foregoing facts it becomes plain that Chloroform is contra-indicated when organic disease of the heart or lungs, and a tendency to apoplexy exist in a marked degree.

Means for resuscitating a Patient when over dosed by Chloroform.—1. Desist from farther administration of the drug.

2. Give the patient an abundance of pure air, by throwing open the windows, using the fan, and sending off as many assistants as can be spared.

3. Dash cold water with some violence upon the body, or pour it from a heigth of several feet.

4. Stimulate the surface, especially over the spine and heart, with hot mustard water, dilute Spirits of Ammonia, mustard plaster saturated with Chloroform.

5. Institute artificial respiration by the Marshall Hall method.

6. Administer injections of turpentine, or pour Chloroform over the scrotum.

7. Apply galvanism in such a way as to stimulate the heart and diaphragm.

8. As soon as the patient can swallow, administer stimulants, commencing with a small quantity and increasing it.

9. Should the patient vomit turn him upon his side and not upon his abdomen—so as not to interfere with the diaphragm—with the head inclining downwards.

Assistants.—When practicable there should be

four assistants, viz: One to administer Chloroform and watch the pulse; a second to compress the artery and apply the Tourniquet if necessary; a third to hold the limb, and retract the muscles; and a fourth to ligate the arteries. If it be difficult to obtain this number, the third assistant can be made to retract and ligate also. It is better to have too many assistants than too few, and the Surgeon should always bear this in mind when making his detail. In field infirmaries, Surgeons will find their duties much lightened by a division of labor, each operating and assisting in turn — the one using the knife having nothing to do with the dressing of the stumps, save to exercise a general supervision over it. These directions are, of course, to apply only when there is a full complement of medical officers present—a rare circumstance in the Confederate service, and a most unfortunate one, as the history of every camp and field attests.— When will wisdom be learned or justice done in connexion with the medical department of the army?

MODES OF OPERATING.—The methods of Amputating are known as the *Circular*, the *Double Flap*, the *Single Flap*, the *Oval*, and the *Diaclastic*. Of these various methods, the *Circular* and *Double Flap* are most in vogue at the present day.

The special ends sought to be attained are to retain enough of the soft parts to cover the bone and to prevent its projecting; to effect as speedy and firm a cicatrization as possible; and to so cover the stump that it may not be liable to excoriate on the least friction. The consecutive treatment has as much to do with the fulfillment of these indications as the choice of methods, and should, consequently, receive the earnest and continued attention of the Surgeon.

Circular Method.—This dates its origin from the times of Celsus, but has been much modified and improved upon by Cheselden, Petit, Bell and Desault. It is performed in two different ways. The method of Desault. Directions: The first incision is carried through the skin and cellular tissue alone, being made by one sweep of the knife, and encircles the limb; then dissect back the skin with a Bistoury for three inches; and turn it over in the form of a cuff (first recommended by Alanson;) then, placing the knife upon the muscles near the fold of the skin, cut through them, by a circular incision, to the bone—taking care to have the edge of the knife slightly turned towards the patient's body. Finally, retract, saw through the bone, ligate the arteries, and bring the wound in apposition.

The Method of Petit.—Directions: The skin being firmly retracted, make the first incision through it and the cellular tissue, by one sweep of the knife, encircling the limb; retract the skin still more and pass the point of the knife under it along the whole extent of its divided surface; cut through the superficial layer of muscles by another circular incision; then retract still more, and divide the deep muscles to the bone. Use the retractor, denude the bone, saw through it, and take up the arteries. The edges of the wound should then be approximated, and the stump treated on general principles.

Double Flap Method.—This was devised by Vermale, and has since been repeated with great success by other Surgeons. The flaps are formed in two ways, either from *without* inwards, by the method of Langenbeck, by drawing the soft parts *off* from the bone, and then carrying the knife obliquely *from* the *surface* and *towards* the bone; or from within outwards, by transfixing the limb with a long, narrow, and sharp pointed knife, at the point of amputation, and then bringing the edge obliquely outwards to form a flap. The same process is repeated for the opposite side, and in this manner double flaps are formed.

The flaps having been thus formed, are held back by the assistants, and all intervening tissues divided to the bone, which is then sawed through. Then take up the arteries, bring the flaps together, apply sutures, and dress the wound.

Single Flap Method.—The origin of this method may properly be referred to Londham, an English Surgeon, who introduced it in 1679. One flap is made in the manner described under the last head and then, the parts on the opposite side of the limb are divided down to the bone, by a semi-circular incision. The flap should be long enough to cover the stump; and, after the arteries have been ligated, should be turned over and secured to the divided surface above, by means of sutures and straps.

This operation is frequently resorted to as a matter of necessity, when the soft parts have been lacerated higher up on one side of the limb than the other, as frequently occurs from gunshot wounds.

The Oblique or Oval method.—This v. as first employed by Langenbeck, and subsequently by Guthrie for the shoulder joint. The incision, by this method, is carried around the Limb in a sloping direction, which is oblique with reference both to the longitudinal axis and the perpendicular diameter of part. The remainder of the operation is performed as in the circular method.

Diaclastic Method of Maisonneuve.—M. Maisonneuve of Paris has proposed a new operation to which he has given the name of Diaclastic, or that by Rupture. According this Surgeon phlebitis or purulent absorption is the accident which most frequently follows amputations, and decides the case unfavorably. It is a matter, therefore, of the greatest possible moment, to resort to someSur gical procedure by which the part can be readily removed, and this fatal symptom avoided. As the surface of a wound after amputation by the knife presents a space open to the action and penetration of the subsequently formed purulent matter, he proposed to divide the tissues by ligatures, or by " instruments, which like scissors bruise the parts during division."

By means of a peculiar contrivance, which it is unnecessary to describe here he fractures the bone and then divides the tissues with an instrument similar to the Ecrasure.‡ M. Maisonneuve, after many trials on the dead subject, has at length attempted the operation on the living; but it can only be regarded as one of the curiosities of Surgical experience.

The more minute manipulations in these different methods will be particularly considered when individual operations are discussed.

GENERAL REMARKS.—Much diversity of opinion exists among Surgeons in regard to the relative advantages of *Circular* and *Flap* operations, each having its zealous advocates, who display much energy and interest in the controversy.

The advantages claimed for the Circular mode are as follows :

1. Cicatrization is more rapidly effected, or the wound heals quicker, while there is less suppuration and sloughing.

2. The Arteries can be more rapidly secured and firmly tied, because evenly divided; while there is no danger of transfixing them, as in Flap operations.

3. The wound can be more readily, efficiently and continuously closed with sutures, so that the water dressings may be employed to greater advantage than where Adhesive Plaster is extensively used.

4. The Vessels contract more firmly, thus to a

‡This is an instrument invented by M. Chaissaignac, the essential part of which is a sort of blunt chain tightened by a screw or by a rack and by pinion. The action of the instrument though slower than that of the knife, is more rapid than that of the ligature, while its influence is direct. It first condenses and then divides the tissues with extreme regularity.

considerable extent obviating the danger of secondary Hemorrhage.

5. Patients can be more safely transported. Macleod affirms that Flaps are knocked about in such a manner as to bruise and injure them severely,—causing sloughing and materially retarding recovery when Patients are carried a long distance either by land or sea.

6. Operations can be performed at a greater distance from the trunk.

The advantages claimed for the Flap operation are:

1. The operation can be more readily and rapidly performed.

2. There is less danger of having the bone uncovered, and of thus exposing the operator to ridicule, and the Patient to additional suffering.

3. The Surgeon is enabled to select a covering for the bone from some of the tissues which remain intact.

4. The muscles can be more readily retracted, and the saw more advantagiously used.

5. The stump is usually better covered, though the work of cicatrization may be delayed. Union by "first intention" frequently takes place in this connexion, the opinions of some to the contrary notwithstanding.

With this statement of the arguments advanced upon both sides of this long mooted question, the Surgeon is left to his own judgment in regard to the the cases which may present themselves, as no specific rule can be given which will apply to each individual injury, and as the best Surgeons vary their operations according to the nature of the circumstances surrounding them.

As a general thing, the Double Flap operation willbe found best adapted to single bones, as the thigh and arm; and the Circular best suited to double bones, as the leg and forearm.

Length of the Flap.—Sir; C. Bell, declares that "the general rule in all cases is to save integument enough to cover the muscle, and muscle enough to cover the Bone, taking care to scrape off none of the Periosteum." This is capital advice and should be regarded. It should also be borne in mind that, after amputations for gun-shot wounds, there is more of tonic muscular contraction than under ordinary circumstances; and hence, greater care should be taken to see that the Bone is properly covered.

Though it is certainly possible to have the flaps of *too great a length*, yet nothing can be more embarrassing to the patient and annoying to the Surgeon than to have them *too short*. The exposure and exfoliation of the Bone follow, as a matter of necessity, and another operation has to be performed, as the only means of correcting the error. A mistake of this kind should be corrected as soon as it is discovered, even before the stump is dressed, by sawing off another portion of the Bone, with an honest acknowledgment of error on the part of the Surgeon. After the expiration of a few days only, it is exceedingly difficult to denude the Bone sufficiently to apply the saw, as it immediately becomes invested with a hard and irregular callus,—defying the knife and rendering its exposure a veritable work of excavation.

Let the Surgeon remember, however, that it is not so much the length of the Flaps which prevents the risk of protrusion of the Bone, but the height at which it (the Bone) is divided above the angle of union of the Flaps.

Varieties of Double Flap Operations.—There are two varieties of this mode of Amputation—viz: When Anterior and Posterior flaps are made, and when covering for the Bone is sought for on either side of the Limb by cutting lateral flaps. To the latter method a serious objection can be urged, even though it is possible to save some blood by cutting the Flap which contains the artery last, the Bone is prone to rise up in the angle between the two Flaps, and, thus, to keep its lower end continually exposed. The same accident may occur when Anterior and Posterior incisions have been made, by turning the Limb upon its side instead of its posterior surface, and thus permitting the muscles to lift the lower extremity of the Bone upwards in the angle between the Flaps, while the Flaps themselves are permitted to fall downwards by the force of their own gravity. These evils can be avoided by proper watchfulness, and their occurrence is consequently a disgrace to the Surgeon. Let him guard against a protrusion of the Bone then, as he values his own reputation, not that such accidents necessarily imply ignorance or carelessness, but as they are thus produced in a large majority of cases, an amount of odium attaches to them and which but few men have the professional *status* to withstand.

Whether the Flap or Circular method be employed, care should be taken to calculate the diameter of the Limb, and to give the skin on either side, at least half that length. The Flaps may even be a little longer on account of the disposition of the skin to retract: and the operation should not be under-taken until the length to be given them is arranged in the mind of the Surgeon, and the precise spot at which the Bone is to be sawn through, accurately determined.

THE POINT AT WHICH AMPUTATIONS SHOULD BE PERFORMED.—The French very properly distinguish (1) the *place* of *necessity*—where there is no choice of site because of the nature of the injury—; and (2) *the place of election* where the most available locality can be selected. The *place of election* varies in different members, though the general rule is to save as much of the Limb as can be done without endangering the patient's life. The facts upon which this rule is based are, the greater utility of long stumps in general, and the diminution of the danger in proportion to distance from the trunk at which the operation is performed. Thus, according to Malgaigne, from 26 Amputations of the smaller toes, 1 death occurred; from 46 Amputations of the great toe, 7 deaths; from 38 partial Amputations of the foot, 9 deaths; from 192 Amputations of the leg, 106 deaths; and from 201 Amputations of the thigh 126 deaths. In the

Crimea the mortality after Amputations of the Thigh was;
for lower third fifty six per cent.
for middle " sixty per cent.
for upper " eighty six per cent.
for Hip one hundred per cent.

The mortality after Amputations of the Arm was:
for the Fore-arm seven per cent.
for the Upper-arm nine-teen per cent.
for the Shoulder-joint thirty-five per cent.

These facts are significant, and should be carefully garnered as the most reliable data upon which the Surgeon can base his judgments, both in Field and Hospital service.

Many Surgeons, and particularly those who prefer the Flap operation, are in the habit of Amputating the Leg at a point about three or four inches below the knee-joint, as the operation can be more conveniently performed there, and as the shorter stump can be more easily managed afterwards. Bockel of Strasburg has however collected the statistics furnished by various authors on this subject, and shown that the mortality attending the higher operation, exceeds that of the lower—the Infra—Malleolar operation,—100 per cent. This he attributes:

1. To the wound being farther from the Body in the lower operation.

2. To the surface of the wound being smaller.

3. To the comparative rare occurrence of Pyæmia and Phlebitis.

When it is possible to obtain artificial limbs of

superior construction, much greater usefulness of the member can be secured after the supra-malleolar amputation has been performed; but, for the attachment of the "wooden-leg," upon which our soldiers must rely under existing circumstances, the shorter stump is far more available.

The rule, enunciated above, is not however of universal application. Amputations through joints are not more dangerous than operations made by section of the bone; and hence, a portion of a member particularly if a small one, can frequently be sacrificed without detriment, to secure the advantages of a disarticulation. Thus a portion of a phalanx may be sacrificed, and the amputation performed at the nearest joint, rather than wait for the saw; and, notwithstanding all the advantages of saving an inch or two of the ulna, and radius, it is better to amputate at the elbow joint than too near it, in order to avoid the disadvantages of the subsequent inflammation. The same remarks will apply to amputations made at a short distance from the shoulder and knee joints,‡ but the same rule does not hold good for the hip joint, as disarticulation there is usually fatal.

Amputations made through the cancellous structures near the ends of the long bones, are less dangerous than those made through the shafts, as they are not so likely to be followed by suppuration and pyæmia.

---

‡Baudens says that his experience in the Crimea assures him that disarticulation of the knee should always be prefered to amputations of the thigh.

MANAGEMENT AFTER AMPUTATION.—As was mentioned before, the success of an Amputation depends as much upon the subsequent managemen- of the case, as upon the manner in which the opera tion is performed. The following rules should govern the Surgeon in this regard.

1. Keep the wound open until the patient has recovered from the shock of the operation or from the effects of the Chloroform, lest Arteries which have been paralyzed thereby, may bleed, and endanger the patient's life.

2. Adjust the Flaps carefully, but not too closely, by means of sutures, and strips of adhesive or isinglass plaster. The sutures should be made of strong saddler's silk (or silver if it can be obtained,) and applied in such a manner as to embrace at least one eighth of an inch of the upper Flap, and one quarter of an inch of the lower.

3. Bring out the Ligatures at one angle of the wound and secure them by a small strip of adhesive Plaster, taking care to handle them lightly and to provide against the possibility of traction during subsequent manipulations.

4. The wound may be dressed in two ways: (1) By inserting sutures at the distance of an inch from each other, supporting them with strips of adhesive plaster, then using a single layer of wet cloth, covered with a waxed cloth to keep in moisture, and applying an iced bladder or water by irrigation. (2.) Applying sutures to the entire length of the wound, drawing the intermediate spaces together by means of Isinglass Plaster, leaving uncovered the angle where the ligatures escape, so

that drainage may be kept up -, and applying the
Maltese Cross by means of a *light* roller, so as to
assist in excluding the air and converting the
wound into a subcutaneous one. No water dressing is used and the stump is left undisturbed.
This mode of dressing is particularly applicable to
Circular Operations, where the skin alone forms
the Flap. Diachylon Plaster is more irritating
and less convenient than Isinglass Plaster, and
should not be used in this connexion, when the
latter can be obtained.

5. It is particularly important to insist upon
absolute rest about the tenth or twelveth day after
the Operation, for at that time the Ligatures are
escaping from the Arteries, and there is danger of
secondary hemorrhage, which is always a dangerous complication.

6. It should not be forgotten that a large majority
of the patients who come under the care of military
Surgeons have been exposed to the debilitating
influences incident to Camp and Hospital life, and
that the demand for nutritious food, stimulants
&c., is unusually great. Without attempting to
decide the much mooted question in regard to the
change of type alleged to have taken place in the
diseases of the present day, it is only necessary to
call attention to the fact that a typhoid tendency
does manifest itself in connexion with the systems
of our soldiers generally, and that the demand, is
usually for a supporting plan of treatment.

7. Apply no bandages after Amputations unless
it be for temporary purposes upon the field, or a
light one to retain the proper dressings.

Bandages have been recommended as a valuable means of arresting muscular contraction, but when it is remembered that the opposing force exerted by such appliances is nothing when compared with the power with which muscles contract when entirely severed,.the fallacy of this proposition is manifest.

Again, they have been employed to prevent involuntary muscular twitching, causing the stump to start, &c. Experience shows and physiology demonstrates the utter impossibility of a sufficient control being exercised by bandages in this regard.

And finally, it is asserted that they prevent purulent absorption, as well as the entrance of air into the veins. This supposes that notwithstanding the pressure of the atmosphere, veins remain patulous after being divided, whereas except, under peculiar circumstances, they immediately close without requiring the intervening agency of bandages applied to the stump. Besides, veins have no such power of suction as is claimed for them under this hypothesis. But they are likewise injurious by complicating the dressings, concealing the stump from view, becoming offensive, and retarding the flow of blood to a part which requires as much of that vitalizing and recuperative fluid as its capabillities will admit. It is better therefore, to support the limb upon a pillow and employ cold water dressings *without bandages*, save such as have been mentioned.

Accidents following Amputation.—The accidents which supervene upon amputations are those which are peculiar to the operation and those which pertain to it in common with wounds generally. The most prominent of those which are peculiar to the operation are: Necrosis of the bone; Conical Stump; Neuralgia of the Stump and Aneurismal enlargement of the Arteries.

The most prominent of those which associate themselves with this operation in common with wounds generally, are: Maggots in the wound, Erysipelas; Gangrene; Pyæmia; Tetanus; and Hemorrhage.

Necrosis of the Bone.—It happens not unfrequently that Necrosis of the Bone takes place after amputation. The remote causes of this accident are: Scrofula, Syphilis, and Cacectic states of the system generally; while the direct or immediate causes are exposure of the bone either by destruction of the periosteum during the operation or inflammation of it afterwards. The signs by which it can be determined are the ordinary symptoms of local inflammation to which are subsequently superadded those which particularly distinguish the progress of that morbid process in bony tissue—such as great pain and swelling, red and glazed condition of the surface, the copious discharge of a very fœted pus, &c., the formation of a sequestrum, &c. The treatment consists in endeavoring to cover the denuded bone, in sustaining the strength of the patient, and in exercising the affected portion.

Conical Stumps.—The bone may protrude in consequence either of the carelessness of the Surgeon in not leaving covering enough, or of the unavoidable retraction of the tissues. The reader is referred to what has already been said in regard to the length of the flaps, and the rules for cutting them.

When it becomes apparent that this accident is likely to occur, the following procedure may be attempted with a reasonable hope of success: Cut a long strip of adhesive plaster two inches and a half in width; apply one end upon the inner side of the limb, beginning if possible, eight inches above the wound; apply the other end upon the other side of the limb in the same manner; make a few turns with a roller wetted or a strip of adhesive plaster, around the limb and over the plaster first applied; to the loop formed by the first strip of adhesive plaster, formed below the amputated surface, attach a small cord; then pass this cord over a small wheel at the foot of the bed, and tie to it a weight sufficiently heavy to bring the soft parts down over the denuded bone. Traction may be kept up in this way for weeks, without inconvenience to the patient, and with the best results. The author recalls in this connexion as illustrative of the advantages of this plan of treatment, the case of Burns of the Louisiana Battalion of Tigers, who was wounded by a conical ball just above the ankle, in a picket fight on the Potomac. Shortly after the first battle of Manassas, he was brought to the General Hospital at Charlottesville, Virginia, and placed under my charge. On enquiry I found that two amputations had been performed on him,—one below the knee joint and the other just above it, the second being necessitated by the protrusion of the bones from the stump. Notwithstanding that the thigh operation had evidently been performed with care, the bone was protruding for more than two inches, while the muscles manifested a disposition to contract still farther. An effort was made to separate the soft parts from the bone and to excise it at a proper distance above the surface of the wound. The bone was found so completely surrounded by a hard and irregular callus, that the work of excavation could not be accomplished, and excision was consequently made on a plane with the divided tissues. The adhesive straps were then applied, as before described and the traction continued for several weeks, at the expiration of which period the bone had been completely and beautifully covered. The principle involved in the treatment of fracture by means of adhesive strips was simply invoked in a new direction and with a satisfactory result.

Neuralgia of the Stump.—It sometimes happens that a distinct tuberose enlargement of the nerves in the stump occurs, attaining the size of a large cherry, and giving great pain by pressing against the bone. Excision of this bulbous extremity is the proper remedy. Again, an important nerve may be included in one of the ligatures, causing pain and paralysis. The stump should be opened and the end of the nerve cut off. In the nervous and hysterical, neuralgic pain frequently occurs, almost defying treatment. As general remedies, tonics, sedatives and alteratives may be administered: while as a topical application the subcutaneous injection of morphia stands unrivalled.

Spasm of the stump must be treated on general principles—tonics, nervous stimulants, bandages to the part, &c., are the most approved remedies. Aneurismal enlargements may possibly be removed by pressure; but if of a more serious character the Artery must be ligated or another amputation attempted.

Maggots in the wound.—This is always a serious and troublesome complication—annoying to the patient and embarrassing to the Surgeon. Their tenacity of life is truly astonishing, while the celerity with which they are produced is truly wonderful. A wounded surface over which a seemingly continuous stream of cold water is flowing, will suddenly and almost miraculously teem with these active and disgusting insects, notwithstanding great care on the part of the attendants. Prevention, however, is every thing in this connexion. If the stream *really* be continuous, and the attention unremitting, the accident cannot occur. When these insects have been developed, their destruction may be secured by either one of the following remedies: calomel, applied in powder, or suspended in water; black wash; creosote and water; an infusion of the marygold; chloroform; elder juice; an infusion of elder leaves and flowers; and various other applications, which it is unnecessary to mention. Calomel and Elder juice are the most reliable, as well as the most harmless of these various remedies.

Erysipelas.—This affection is connected with some depraved and altered condition of the blood, particularly of the red corpuscles, and is essentially the local manifestation of a general or constitutional malady. It is really a disease of debility in as much as the nervous centres are not supplied with their normal and necessary amount of healthy pabulum, in consequence the precedent changes in their vitalizing fluid. The pathological conditions which characterize this disease may be thus expressed.

1. Changes in the blood, whereby the Corpuscles are rendered less stimulant and nutritious to the nerve centres, &c.

2. Changes in the nerve centres resulting from the absence of their necessary food, whereby they become irritated and not duly stimulated.

3. In consequence of this irritation the whole system, that machinery of which the centres are the motive power—acts irregularly; and hence *fever, local congestions and inflammations,* disturbance of the secerning organs, &c, ensue.

4. As a result of this want of stimulation, the centres lose their tone,—their generating power abates, and the whole system becomes decidedly debilitated. The symptoms which distinguish Erysipelas are so well known as scarcely to require enumeration. They are the following: a reddish flush rapidly spreading over the surface; a peculiar stinging and burning pain; considerable swelling; much tension; tenderness on pressure; great heat; and tendency to effusion, together with a full, frequent, but weak pulse; a dirty and coated tongue; and deranged gastro-intestinal secretion. There are two principal varieties, viz: the simple Cutaneous and the Cellulo-Cutaneous or Phlegmonous. The former limits itself to the skin, while the latter extends to the cellular tissue which separates that

tissue from the muscles and the muscles from each other. The phlegmonous is far the more serious and fatal affection. Its symtoms are more violent *ab initio*, while a tendency to rapid and extensive suppuration is one of its most serious characteristics. Beneath its hurried and fearful footsteps muscles are uncovered, blood vessels exposed,' bones robbed of their covering, joints opened, and whole members terribly and completely devasted. In its train comes Hectic with all its frightful retainers, the ghastly herald of an early death.

When suppuration has been established, and pus is evacuated externally, openings are formed bounded by edges of mortified cellular tissue, and cicatrization takes place most tardily if at all. In many instances sloughs of great extent are produced, while the fever continues, the general disturbance augments, an intestinal inflammation is excited, prostration ensues, and a fatal diarrhœa is developed. Occasionally Erysipelas is primarily and essentially gangrenous, marching with rapid strides to a fatal termination, and utterly defying the skill of the Surgeon. Researches into the pathological anatomy of this affection clearly establish that the inflammation incident to it affects in different degrees the skin, the tegumentary vessels, the cellular tissue and the lymphatic system, and that its fatality and violence are in proportion to the depth and number of the structures involved.

Without discussing at length the question of the contagiousness of Eresipelas, it is sufficient to remark that though this character has been claimed for it by Lawrence, Arnold, Willan and Erichsen, a large majority of modern pathologists totally and emphatically repudiate the idea. The case referred to by Erichsen and quoted by Chisolm as illustrative of its contagiousness, is not sufficiently conclusive as the appearance of the disease, under the circumstances alluded to, may have been a mere coincidence, and as proof equally as strong can be adduced in support of the communicability of any disease. If the disease be strictly contagious, in the ordinary acceptance of that term, why were only *the sick* attacked, while the physicians, nurses and attendants escaped? It may also be asked in the same connexion, why it does not "spread" in all Hospitals alike, or communicate itself to those who dwell in their immediate vicinity, and who are in communication with patients suffering from the disease?

The circumstances which associate themselves with the development of Erysipelas are:

1. A system which has been debilitated by previous exposure, fatigue, loss of blood, indulgence in venery and intemperance, or improper food.

2. Certain special hygienic conditions,—such as deprive the patient of those surroundings which are essential to the preservation of his system in its normal *status*.--Among these are impure air, want of cleanliness, non-nutritious food, abnormal electrical conditions of of the atmosphere, &c.

These circumstances may so combine as to develop the disease spontaneously, or they may furnish those conditions which necessitate its dissemination when the morbific elements are furnished by a case already existing, as occurred in the instance alluded to by Erichsen; but it cannot be pretended that without these particular conditions—this special preparation of the system for the invasion of the malady—erysipelas can be propagated by contract. By a " contagious disease," according to the teachings of the ablest writers, is meant an affection which under *ordinary* circumstances, and with the human system in its *normal* condition, attacks a large majority of those who are brought in contact with it. Erysipelas does not thus propagate itself save under *extraordinary* circumstances, and when the normal *status* of the system has been materially altered; and hence *à priori*, it is not a "contagious disease" according to the usual acceptation of that term. A malady which is not "contagious" cannot be "infectious," inasmuch as the latter term implies propagation by *contact* with some emanation—ærial or gaseous it may be—from an affected system.

The treatment of erysipelas has become far more rational and successsful within a few years. The seeming violence of the febrile phenomena, is no longer regarded as an indication for antiphlogistic remedies; but, regarding it as essentially and exclusively a disease of debility, the profession has learned to depend upon tonics and stimulants, as the agents best calculated to stay its rapid and fearful progress.

So, likewise, the doctrines of Higginbottom, in regard to the pathology of the disease, have been overturned by the more enlightened views of Chomel, Blanche and Biett, and it is now recognized and treated as a constitutional disease which expresses itself in a topical inflammation.

The primary indications are to administer remedies which will restore the altered corpuscles to their original purity, and at the same time give tone and power to the exhausted nerve centres. Muriated Tincture of Iron and Sulphate of Quinine are the remedies which will most successfully accomplish these results; and the following prescription will best combine them:

℞ Tinct: Muriatici Ferri, ʒ ij.
Quiniæ Sulph: Ɔ ij.
Aqua, font ʒ iv

M.
S.—A teaspoonful every third hour.

The Muriated Tincture of Iron is not only more rapidly absorbed, but it also possesses the power of restoring the altered corpuscles and of controlling the local inflammation, by constringing the capillaries of the affected structures. The Quinine acts directly upon the the nerve centres increasing their capacity for the production of that subtle nervous influence upon which the integrity of the whole organism depends.

Stimulants should also be employed for the purpose of giving tone and strength to the exhausted system; while liquid and nutritious food constitute a necessary and most important addition to the treatment.

Should the progress of the disease be complicated with gastric disturbance, an emetic or a mild purgative should be administered, for the purpose of removing all offending matters, and of restoring the secretions, but not with reference to its depletory effects.

Local applications are neither to be despised nor too much relied upon. Mercurial ointment, as recommended by Ricord, the Camphor Water of Malgaigne, Velpeau's Ointment of Sulph: Iron, Nitrate of Silver as proposed by Higginbottom, fomentations of elder flowers, poppy heads, cranberries, hops, &c., solutions of nitrate of potash, sugar of lead, carbonate of soda, and chlorate of potash-creosote, collodion, ice, tincture of lobelia, dilute acetic acid, white lead, muriated tincture of iron olive oil, &c, &c., have all been tried, and have their admirers. The great remedy however is *Cold Water* medicated according to the indication, and applied in conformity with the rules and principles enunciated in the first chapter of this work. Scarifications, both as a means of relieving the local hyperæmia and of permitting the escape of pus are invaluable.

It, of course, becomes a matter of the greatest moment to improve the sanitary condition both of the patient himself, and of those with whom he may be associated. It is well, therefore, to place the patient in a tent in the open air, removed from his companions, and so situated that he may get an abundant supply of fresh air. His body should at the same time be kept clean, and an abundant supply of good food furnished.

The greatest care should be immediately taken to ventilate, and purify the Hospital, to see that wounds are frequently and properly dressed, and their products removed, to have all vessels cleansed as soon as they are used, to keep the bed linen fresh and clean, to empty the spit boxes regularly, to provide pure and nutritious food, to administer stimulants freely, to cheer and console the patients, to segregate the sick and wounded, to fill up sinks and change the location of privies, and to do such other things as the laws of health require. An Epidemic of Erysipelas is not likely to prevail if the hygienic conditions are good, and when they are so, the separation of sporadic cases no longer becomes a necessity.

Pyæmia—By this term is meant that pus poisoning which sometimes takes place in connexion with wounds produced by the amputating knife and other causes. It is preceded by a stage of incubation in which the patient is restless, sleepless, uncomfortable, feverish, pale, and apprehensive of evil. The disease proper is ushered in by violent rigors, which continue to occur at regular or irregular intervals, followed by high fever, jaundiced hue of the skin and conjunctiva, a furred tongue, a frequent but feeble pulse, delirium or coma, gastric irritation, diarrhœa, sardonic countenance, great thirst, copious sweats, and extreme restlessness. The patient gradually grows more feeble, the joints inflame and swell, the organs generally show greater signs of disturbance, the pulse sinks, collections of pus occur in the various tissues, and the wound frequently becomes boggy and yielding but comparatively dry, and death finally closes the scene.

These symptoms are manifestly due to the presence of pus in the blood—poisoning that fluid and acting as an irritant to the tissues and organs generally. The purulent fluid is introduced into the circulation in two ways, which are entirely distinct from each other—viz:

1. By inflammation and finally suppuration of the internal coats of the Veins.

2. By the absorption of pus,—modified but not improved—into the blood.

The Blood, as previously stated, is poisoned by the presence of this product to such an extent, that it not only fails to supply pabulum to the tissues, but becomes positively irritating to them. There are consequently developed in various parts of the body points of irritation at which the blood accumulates until inflammation is developed, and an effusion of lymph takes place. This lymph being but poorly organized because of the blight impressed upon the whole mass of the blood, readily and rapidly breaks down into pus; and hence those multiple abscesses are formed which constitute the most prominent feature of the disease.

Pyæmia being essentially a disease of debility, requires to be treated by tonics, stimulants, and a nutritious diet. The restlessness and insomnia must be controlled with opium; the diarrhœa with opium, and astringents combined; the inflammation of the veins combatted on general principles and the purulent matter given a free vent. Macleod suggests the propriety of ligating the chief vein with the artery, as a means of cutting off the channel by which the poison is conveyed into the system. This may prevent the absorption of pus, but it is likely to defeat its own object by inducing phlebitis. Amputations after the development of Pyæmia, so far as the authors observations go, are invariably fatal. The application of Chlorinated washes to the surface of the wound will be found useful, in conjunction with the general treatment marked out. All that was said in regard to separating the patients affected with this malady, improving general and individual hygiene, &c., in connexion with Erysipelas, applies with equal force to Pyæmia.

Hospital Gangrene.—This is also a disease of debility, and results from the influence of a blood poison acting on an enfeebled constitution. It is both contagious and infectious. The symptoms which characterize it are: feverishness, loss of appetite, sleeplessness, coated tongue and deranged bowels, followed by a dry and painful condition of the wound, the appearance of an ash colored—slough, which is soft and pulpy, engorgement of the neighbouring skin, eversion and undermining of the edges of the wound—which are of a livid red color—, and finally the complete breaking down of the dying tissue, with the development of a thick and dirty fluid, and a peculiarly offensive odour. The mortification extends rapidly and the system sinks under its baneful influence. The treatment consists in sustaining the system with tonics and stimulants, and destroying the poisonous ichor, from which the local and general poisoning results. The first indication is accomplished by the free use of Quinine, Iron, and Brandy; while the second is fulfilled by such remedies as the actual cautery, caustic potash, nitrate of silver, tincture of iodine, creosote, chloride o iron, lemon juice, pyroligneous acid, nitric acid, muriatic acid, &c., followed by irrigation. To allay pain, calm nervous disturbance, insure sleep, &c.—Opium may be freely used. But above all things *remove the patient from the infected atmosphere;* and surround him with those things which hygiene and humanity demand for his health and comfort.

Tetanus.—Tetanus is a peculiar condition of the nerves centres, characterized by the following phenomena; the wound is dry and painful; the patient shows signs of mental agitation and fright; convulsive movements of the face, and of the members, particularly of the arms, take place; deglutition and mastication are rendered difficult— preceded by soreness of the throat, and followed by locking of the jaws; contraction of the muscles of the neck take place; the abdominal muscles become hard and knotted; violent and repeated spasms occur, while the pulse grows feeble, the countenance sardonic, and the skin profusely moist. Tetanus is said to be *complete* when all the muscles of animal life are equally and thoroughly contracted. Under these circumstances, the body becomes so thoroughly stiffened as to seem all of one piece, and will break rather than bend—the fingers however are an exception and still remain flexible. The face especially is remarkably fixed and motionless, and wears an expression which resembles that of death or of mortal agony. The pain of this affection is terribly **severe**, being similar to that produced by *cramp* of *the muscles*, such as every one is familiar with. The intellectual faculties remain intact up to within a brief interval before the approach of death. The appetite is good, but the impossibility of deglutition frequently produces death by starvation, according to Larrey.

Tetanus is a very grave malady nearly always; terminating fatally, especially when of traumatic origin, and involving all the muscles of animal life.

According to the muscles involved, it is styled trismus, emprosthotonos, opisthotonos and pleuristhotonos.

It is essentially a lesion of the nerves, but as yet pathology has not ascertained its essential nature, notwithstanding the recondite researches of Bouillaud, Begin, Andral and Magendie.

The theraputics of tetanus are no better settled than its pathology. Being manifestly a disease of debility, it is a matter of the first importance to sustain the patient, which has to be accomplished by means of enemata, on account of the difficulty of deglutition.

Cruveilhier regarding asphyxia as the usual mode of death, in consequence of the convulsive action of the respiratory muscles, proposed to prevent this fatal result by inducing violent but voluntary movement of the same muscles. He compelled his patients to make forced and profound inspirations until the contractions were overcome.

Busse recommends friction with alcoholic tincture of belladonna, particularly over those points where the convulsive rigidity is greatest.

Larrey cut short the disease in its forming stage by amputating the limb or dividing the nerve.

Fournier treated the disease successfully by means of sulphurous baths, and Paré testifies to the efficiency of the same practice.

Chloroform has many advocates; extract of Cannabis indica, opium, belladonna, woorara, and in fact nearly all the remedies in the Pharmacopœa have been recommended and successfully used. No specific remedies can be relied on, but the following is perhaps the best plan of treatment:

Empty the bowels, by means of scammony, aloes, gamboge or croton oil; divide the principal nerve of the part; apply warm water medicated with opium to the wound; administer chloroform freely until anæsthesia is induced; and inject morphia subcutaneously, either immediately over the track of the principle or in close proximity to the nervous centres which seem most involved. If the strength of the patient can be supported there is some prospect of a favorable termination. Sleep is absolutely necessary to the comfort and salvation of the patient. Nothing can be more important than to remove all foreign bodies from the track of the wound, and to use such remedies as are calculated to relieve the local inflammation.

The patient should be made as comfortable as possible, quiet enjoyed, and an *equable temperature* preserved. The experience of all Surgeons establishes the fact that changes of temperature are prolific sources of this disease—a circumstance which should be remembered both as a means of preventing and curing tetanus.

Tetanus is either traumatic or idiopathic according to the mode of its production.

Hemorrhage.—As the subject of Hemorrhage is treated of at length in a separate chapter, it is not necessary to consider it here.

Sutures—In General.—As soon as hemorrhage has been arrested whether the wound be accidental or artificial, their edges should be brought together for the purpose of securing a speedy union between them, and retained in position by means of *sutures*, together with adhesive straps.

The following general rules for the application of Sutures should be studied and observed.

1. Enter the needle at an angle of 65°, at a distance from the margin proportionate to the length of the wound and its tendency to gap.

2. Have the points of perforation on each side and the amount of tissue embraced in both margins of sufficient extent to close the parts without wrinkling.

3. If the sutures alone are relied upon to close the wound, a sufficient number must be used to make the line of union complete. As a general rule it is better to employ a smaller number of sutures and to bring the parts more completely together with adhesive strips.

4. The knots of the Ligatures must be arranged upon the upper side of the wound,—so as not to be affected by the discharges from it—and tied only moderately tight.

5. Sutures should be removed as soon as adhesion has taken place,—not all at once, but separately and carefully.

6. In the application of sutures avoid wounding nerves, vessels, serous membranes or tendons.

7. Should union take place by "first intention," the sutures may be removed about the eighth day, but if by "second intention," not under a month.

PARTICULAR SUTURES.—It may be well before advancing farther, to consider briefly the different varieties of sutures employed in Surgery.

1. *The Interrupted Suture.*—This is formed by passing a needle and thread, through the skin and subcutaneous cellular tissue from *without inwards* on one side, and from *within, outwards* on the other, fastening the ends of the thread together, and cutting them off close to the wound.

The stitches are proportioned in number to the extent of the wound, and are usually inserted at the distance of an inch from each other. This is the form by which the margins of wounds made in performing amputations, are kept together with the assistance of adhesive straps.

2. *Glover's Suture.*—This differs from the last in that the edges of the wound are brought together, and the needle and thread passed at once through both of them, then brought over to the same side, and passed, again through both edges, and so on to the end of the wound, making at each stitch a loop which is drawn tight—precisely as the edges of a glove are "whipped" together. This is not much employed at the present time.

3. The *Quilled Suture.*—This is applied in the same way as the interrupted Suture, only the needles

are armed with double threads, so that one of the extremities forms a loop. All the stitches being made on the same line, a peice of quill is passed through the loops, and the threads on the other side of the wound are separated and tied over a similar bit of quill, with sufficient force to bring the sides of the wound together, and to keep them there.

4. *Twisted Suture.*—A round, and straight needle of gold or silver—or a common pin is pushed through the edges of the wound, from *without inwards* on one side, and from *within outwards* on the other. The first needle being thus introduced, a thread is passed under it on either side, and sufficient force exerted to bring the edges well together: another is similarly placed, and a third, or as many as are wanted. Then taking the ends of the thread they are crossed in front of the first needle, and brought again under its extremities, so as to form a figure of eight, repeated four or five times. They are then passed under the second, and similarly twisted, and so on for every needle introduced. When the last turn has been made, the two ends are tied together in a knot or bow. A small compress of lint should be placed under the point of each needle to prevent it from irritating the skin.

Percy long since recommended lead as a good material for Sutures, but experience has demonstrated that silver wire is incomparably the best material for such a purpose, in as much as it is both exceedingly ductile and particularly non irritating to the animal tissues. Physic used kid skin rolled into small cords. Dr. Simpson of Edinburg, recommends wire made of gold, platinum or copper as a substitute for the ordinary suture. Dr. Eve has employed, with the same end, fibres from the sinews of the deer. When they are to remain in position for a long period, employ Metallic Sutures, particularly those made of silver wire, but under ordinary circumstances, it is best to use those composed of some organic material. Thread made of silk, flax or cotton can generally be found by Surgeons every where, and when properly waxed, possess sufficient pliancy and strength for all practical purposes.

The Glover's Suture generally puckers the wound and may be replaced in many instances by the Interrupted; the Quilled Suture causes the bottom of the wound to unite while its edges remain open ; and the Twisted Suture by compressing the flesh only at certain points is more liable to cut through, and disengage itself prematurely than the others. Experience has shown that Sutures made of animal tissue do not possess any decided advantages, and they have fallen into disuse.

CHAPTER III.

PARTICULAR AMPUTATIONS.

Having given the general rules which govern Amputations, it now becomes important to describe in detail the methods of procedure in individual operations.

AMPUTATIONS OF THE LOWER EXTREMITIES.—Under this head are included Amputations of the Foot, Ankle-Joint, Leg, Knee, Thigh, and Hip-Joint.

Amputation of the Toes.—Directions: Seize the phalanx firmly and bend it so as to give prominence to the joint; make an incison across the joint, cutting well into it; divide the ligaments carefully on either side; and then carry the knife through the joint and cut a flap from the under surface of the toe. The flap should then be brought over the surface of the disarticulated joint and secured by ligatures or adhesive plaster. The toes may be amputated at the second joint in precisely the same manner.

Amputation of the Great Toe at the Meta-tarsal Articulation.—Directions: Pass a narrow bistoury up on one side of the proximal phalanx as high as the articulation; carry it then across the joint, turning the point so as to cut the ligaments and open the articulation; lay the blade flat against the toe and cut out a flap on the opposite side. The joint really lies much deeper than one unacquainted with the Anatomy of the part would suppose,—a fact which should be remembered in introducing the knife. It is important to preserve the distal end of the meta-tarsal bone so as to strengthen the foot and prevent lameness. The *Meta-tarsal* bone can be removed by an operation similar to the last,—the first cut being extended to the tarso—meta-tarso articulation. Avoid the anterior tibial artery in opening the joint, for it dips near this point between the meta-tarsal bones. There is always danger of lameness and the operation should be avoided, if possible.

Amputation of all the Toes at their Meta-tarsal joints.—Directions: Make a transverse incision along the dorsal aspect of the meta-tarsal bones; divide the tendons and lateral ligaments of each joint in succession; dislocate the phalanges upwards; and then, placing the knife between the meta-tarsal extremities, cut a flap from the skin on the plantar surface, sufficient to cover the heads of the exposed bones. Tie the arteries; bring the flap in position; and lay the foot on its outer side so as to facilitate the discharge of pus.

Amputation of all the Meta-tarsal Bones.—Directions: Find the point at which the great toe articulates with the inner cuneiform bone; make a semilunar incision beginning at the projection of the scaphoid and terminating at the outer side of the tuberosity of the fifth meta-tarsal bone; turn the small flap thus formed backward, pass the knife around and behind the projection of the fifth meta-

tarsal bone, so as to divide the ligaments which
connect it with the cuboid; depress the toes and
cut the remaining ligaments; disarticulate the
third and fourth meta-tarsal bones; then attack
the first meta-tarsal and finally the second, which
being locked between the three cuneiform bones
is difficult to dislodge and should be sawn across
in some instances. All five bones being detached,
carry the knife behind them, and cut a flap from the
sole of the foot of sufficient length to cover the exposed surfaces of the disarticulated bones. Ligate
the Arteries; bring the flaps in position, and keep
the foot slightly elevated. This is known as
Lisfranc's Operation.

Instead of disarticulating, the bones may be
sawn across, a little in advance of the articulation
as proposed by Hey and Cloquet, facilitating the
operation, and giving results equally as satisfactory.

Amputation through the Tarsus.—Directions.—
Find the the joint at which the cuboid articulates
with the os-calcis, and the point at which the
scaphoid articulates with the astragalus; make a
semilunar incision across the front of the foot connecting these two points; turn back the anterior
flap, and divide the ligaments which connect the
four bones mentioned above; then pass the knife
through the joint, and cut a long flap from the
sole of the foot. Tie the dorsal and two planter
arteries; round off the extremity of the flap before bringing it into position; and take special
care during the cure, to have the gastrocnemial
muscles completely relaxed, by keeping the foot

upon its outer surface over a pillow. This is known as Chopart's operation.

Amputation at the Ankle Joint.—To Syme belongs the credit of having elevated this operation to its proper position in the Surgery of the present day. Being less dangerous than amputation of the leg, and particularly successful as to its results both in America and Europe, it is now regarded with great favor by the profession.

Directions.—Make a curved incision across the instep, from one malleolus to the other; make a a second across the sole of the foot; dissect up the flaps and expose the joint; disarticulate the os calcis and astragalus with the rest of the foot; and then remove the projections of the malleolar process either with the saw or forceps. Should the joint itself be involved, a slice of the lower end of the tibia and fibula may also be removed.

It is difficult to dissect the flap at the heel, and particular care should be taken not to cut through it, or to wound the posterior tibial artery.

Syme's operation is a great improvement on the older methods of disarticulating at the ankle joint.

The operation of Syme has been modified by Pyrogoff, by retaining a portion of the Calcaneum, and thus imparting greater length and rotundity to the Stump. Directions.—Make a curvilinear incision around the foot in front; make a second incision under the sole, extending from the front of one malleolus to the other; dissect up the flap, divide the different ligaments and detach the astragalus; apply the saw just behind the astragalus and divide the anterior portion of the calcaneum; remove the malleolar projections together with a thin layer of the extremity of the tibia; tie the vessels and bring the flaps accurately together. The advantages of this operation are that a larger and better stump is secured, there is less danger of wounding the posterior tibial artery, and the posterior flap is not so liable to form a pouch for the accumulation of pus.

Remarks. — These operations are undertaken with a view of saving as much of the foot as possible, in order that greater support and a more convenient stump may be secured; and though in proper hands they constitute the most valuable and scientific of Surgical measures, they should never be undertaken without a knowledge of the anatomy of the parts, and an acquaintance with the rules which have jurst been enunciated. Either Lisfranc's or Hey's operation is preferable to Chopart's or Syme's, when admissible, in consequence of affording a greater length of Foot and securing a less tender stump, Pancoast's proposition to sever the Tendo Achillis, is a good one. Syme's operation does better for chronic diseases of the foot than when made for traumatic lesions. Pyrogoff's modification is liable to the same criticism, though a beautiful operation in itself. Malgaigne in his statistics of amputations in the Hospitals of Paris found the mortality after the removal of the great toe one to six; and after the removal of the smaller toes one to twenty-six; while in amputations of the foot the proportion of deaths was twenty-five per cent. For the statistics of the operations performed upon the foot in Richmond the reader is referred to Table "A" of the appendix.

Amputation of the Leg.—Directions for the Circular Operation: Administer Chloroform; bring the Patient down upon the Table; command the Artery either by applying the Tourniquet to the Femoral Artery or compressing that vessel against the Pubes; have the limb well supported and the skin drawn upwards; make an incision through the integuments entirely around the Leg; dissect up the integuments for about two inches or two inches and a half and turn the cuff back; and then divide the muscles down to the Bone. This being done, pass a double catline between the Bones, so as to divide the interosseous membrane; and then having drawn the muscles back by means of a three tailed Retracter, saw through the tibia and fibula—engaging it in the larger Bone firmly but completing the section of the smaller one in advance of the other.

Smooth off the bony surfaces with the nippers; ligate the Anterior and Posterior Tibial Arteries, and such branches as may require it; and dress the stump according to the directions previously given.

Directions for the Flap Operation.—Place the Patient in position; administer chloroform; bring the limb down until it projects well over the lower edge of the table; ascertain the exact locality of the bones and transfix the limb by passing the knife horizontally behind them, and not between them; cut a flap from the posterior muscles about five inches in length; and, then, make a semilunar incision across the anterior face of the Leg, connecting the two points at which the point of the knife was made to *enter* and to *leave* the Limb. After this, dissect back the anterior Flap slightly; divide the interosseous muscle and ligament with a double edged catline; use the retractor; and saw through the bones, as previously directed. Mr. Furgusson directs that the Anterior Flap be made first, by placing the Heel of the Knife on the side of the Limb most remote from the Surgeon, and then drawing it across in front of the Limb. As soon as the point of the Knife arrives at the opposite side, the limb must be transfixed and the posterior Flap made as above described.

See that the bones are of equal length and that their edges are smoothed off; bring the edges accurately and evenly together; and remove the spine of the tibia, if it project too much.

Remarks.—The Flap Operation may be performed at any point above the Ankle Joint, where a posterior flap can be obtained, though the rules mentioned above, should always be borne in mind. Great care should be taken not to push the knife between the bones or transfix the main artery; never operate above the tuberosity of the tibia lest the joint be opened or injured by the subsequent inflammation; and do not forget to shorten the Tendo Achillis when the operation is performed near the Ankle.

STATISTICS.

Pennsylvania	Hospital	69	operations, mortality	42	per cent.
New York	"	102	"	"	84 " "
Massachusetts	"	23	"	"	21.7 " "
Reported by Malgaigne	"	192	"	"	58 " "
University College		14	"	"	14 " "
Reported by Macleod		101	"	"	30.3 " "
" by Sorrell		123	"	"	43.9 " "

The reader is referred to Table "B" of the Appendix for fuller information on this subject.

Amputations at the Knee Joint.—Amputations at this joint may be performed in two ways:

1. By making a *large anterior* flap of skin and a *short posterior* one of muscle.

2. By making a short flap of skin in *front* and relying upon the gastrocnemius *behind* for a flap of sufficient length to cover the joint.

Directions for process No 1.—Make an eliptical incision upon the anterior and lateral surface of the limb, from the centre of one condyle of the femur to the same point on the other condyle. This incision must have its convexity downwards, and should embrace surface enough to cover the joint after it has been exposed. Dissect up the flap of skin just made; divide anterior ligaments and open the joint; then divide the lateral and posterior ligaments; and, finally, carry the knife behind the joint and cut downwards and backwards, making a short flap from the muscular tissue on the posterior aspect of the leg. This being done, retract and saw off a portion of the condyles of the femur, so as to secure a smooth surface over which to adjust the flaps, &c. The patella should not be removed in the operation.

Directions for process No. 2.—Between the same points, i. e., the centre of the condyles make an eliptical incision embracing but a short flap, upon the anterior and lateral surfaces of the limb; divide the ligaments as before; and finally pass the knife behind the disarticulated extremities, and cut downwards and backwards, making a sufficient flap of the muscular tissue behind to cover the exposed surface. Then retract, and saw off condyles as before. The Popliteal with its branches, the inferior articular, middle articular, and gamellar, is cut in this operation, and should be immediately tied. The wound is closed and treated in the usual manner.

There are other methods of performing this operation, but the plans proposed above, will be found to answer sufficiently well for all practical purposes.

Remarks.—The propriety of amputating at this joint has been much questioned by Surgeons, yet Velpeau, Baudens, Pancoast, Malgaigne and Macleod have all declared in favor of it; and it may be resorted to in connexion both with primary and secondary operations. It can be performed very expeditiously, but, there, is danger of subsequent inflammation, and perhaps of a tardy convalescence. The advantages which this amputation possesses over that of the femur may be thus summed up:

1. The shock to the system is less.

2. A larger and more available stump is secured, while it is less liable to ulceration.

3. A false leg can be more readily attached, and greater power is obtained for progression.

4. The Medullary Canal is not interfered with, and

the extremity of the femur being well supplied with Blood Vessels, there is less danger of exfoliation.

5. The Artery being in the centre of the Flap and but few ligations being required in the operation, there is less danger of hemorrhage.

6. The operation is not so fatal as that for the fermur.

STATISTICS.

Reported by Macleod,	Operations	8	mortality	50.	per cent.
Reported by Smith,	"	86	"	43	" "
Reported by Malgaigne,	"	9	"	77,	" "
Reported by Pager,	"	27	"	40	" "
Reported by Sorrell.	"	2	"	00	" "

Mr. Baudens affirms that his experience in the Crimea convinces him that disarticulation of the Knee ought always to be preferred to amputation of the thigh, and in this opinion he is sustained by Macleod and Malgaigne, as well as many other Surgeons of the highest character and widest experience.

Amputation of the Thigh.—Directions for the Circular Operation.—Place the patient upon the amputating table; administer chloroform, and bring him down upon the table until the wounded leg projects well beyond its lower margin—being supported at the knee by an assistant. Compress the femoral artery either by tourniquet or digital compression against ramus of pubes; direct the assistant to seize the limb with both hands just above the point selected for amputation, and to draw the skin forcibly back; grasp the limb with left hand so as to steady it; carry the hand under the thigh, and make an incision at one sweep completely round the limb through the fat down to the fascia, dissect up the skin, &c., about two inches and a half; and having turned the cuff back, with one circular sweep of the knife, divide the muscles down to the bone. This being done, separate the muscles from the bone for the space of about an inch; divide the periosteum; retract the flaps; and saw through the bone. In cutting through the muscles, it is a matter of importance to turn the edge of the knife *towards* the body of the patient so as to make a more conical flap. Instead of dissecting up the skin, and turning it back, some Surgeons simply retract forcibly after the first incision has been made, and, then, with the edge of the knife turned towards the patient's body, and the retraction continued, cut through each successive layer of muscular tissue until the bone is reached, when the periosteum is divided and the bone sawed through, as described above. The arteries should then be ligated,—the femoral first, and the profunda next, if it be cut, and all pressure *suddenly* removed, so as to encourage hemorrhage from any vessel that may have been overlooked. After all this has been done bring the cut edges together after the manner already pointed out, and dress the stump according to the circumstances of the case, upon the plans already discussed.

Directions for Double Flap Operation.—Administer Chloroform; arrange the patient upon the table; compress the main artery as directed in the last operation; and, then, having transfixed the limb, by passing the knife *in front* of the bone, and as near to it as possible, cut a flap, of about five inches in length, from the *anterior* portion of the thigh. This being accomplished, insert the knife in the upper portion of the wound depress it and transfix the limb, by passing the instrument *behind* the bone, and, then cut a flap from the muscles which cover the *posterior* portion of the thigh, a little longer than the anterior flap. Turn these flaps back; and, having cut through all intervening tissue, divide the Periosteum, use the retractor, and then saw through the bone. Ligate the Arteries, bring the flaps together, and dress the wound.

The patient should then be removed to bed, and the stump supported on a pillow,—if he has been brought to the Hospital. Should the operation be peformed at a field infirmary, see that he has recovered thoroughly from the shock of the operation, and the effects of the chloroform, and then, move him to some comfortable position, supporting the stump with a knapsack, or whatever may be convenient.

As before remarked, flaps may be made from the *inner* and *outer* side of the limb, by transfixing it from above downwards; but this operation is objectionable for the reason that the bone tilts into the upper angle if the wound remains uncovered.

Remarks.—There are some general rules which should be remembered in this connexion.

1. Stand on the *outer* side for the *left* leg, and on the *inner* for the *right*—always be prepared to grasp the limb with the left hand, above the amputating point.

2. Arrest hemorrhage from large veins by elevating the stump, and compressing with the finger;

but should this fail, they must be ligated. Oozing from the bone may be arrested by holding a compress firmly against it for some time.

3. Restrain obstinate oozing from small vessels.

4. Divide the bone evenly, and use the bone nippers to render its surface smooth and less irritating.

5. Be careful not to include nerves in the ligatures applied to the arteries.

6. Do not pull at the ligatures until about the 12th day, lest secondary hemorrhage ensue.

7. The Surgeon should remember, as a cardinal principle, that the lower an operation is performed upon the thigh, the *greater* are the chances for the patient's recovery, and *vice versa*. He should, in fact, regard, not only every inch, but *every line*, saved, as securing a most material advantage,—an advantage which cannot be denied to the object of his care and skill, without jeopardizing, to a still greater degree, the life of the patient.

8. All experience demonstrates the advantage of primary amputations of the thigh; and the absence of regular appliances, proper means of transportation, &c., upon the battle field, should not afford a sufficient inducement to postpone the operation—whatever the rank or claims of the sufferer. Decide judiciously, and operate speedily, if you wish to save the life of the patient. Delay is death—slow, it may be, but often inevitable.

STATISTICS.

Reported from University College, Operations 19 mortality 58 pr. ct.
" by Malgaigne, " 46 " 75 "
" " South, " 24 " 100 "
" " Buel, " 34 " 59.16 "
" " Norris, " 4 " 00 "
" " Macleod, " 164 " 64 "
" " Mounier, " 46 " 82.6 "
" " Sorrell, " 172 " 59.8 "
" " Sedillott, " " 87.5 "
" " Esmarch, " " 60 "
" " Baundens, " " 51 "
" " Alcock, " " 60 "

Macleod after giving a very large number of cases calculates that the average mortality of Primary Amputations for gunshot wounds alone, is 65.5 per cent; and for secondary operations the mortality is 79.0 per cent. In civil Hospitals, the mortality, according to the tables furnished by that author, is for Primary Amputations 69.6, and for secondary 75.4. The per centage of mortality for secondary Amputations after the first battle of Manassas,—and but few Primary operations were performed—greatly exceeded this; while the relative success of the two varieties of Amputation as indicated by operations performed in connexion with the Richmond Battles is much more decidedly in favor of the Primary than is established by the records of other fields.

From the statistics on file in the Surgeon General's office, it appears that there were performed in and around Richmond, from June 1st to August 1st, 1862, 70 Primary Amputations upon the thigh, of which 16 were circular and 10 flap, and 44 not stated, with a mortality of 36.9 per cent,—or 56.2 for the circular operations, 30 for the flap, and 31.8 for those not stated; 61 intermediate amputations, of which 9 were circular, 6 flap, and 46 not stated, with a mortality of 80 per cent,—or 66.6 for the circular, 83.3 for the flap, and 82.4 for those not stated; and 41 secondary amputations, of which 7 were circular, 2 flap, and 32 not stated, with a mortality of 88 per cent,—or 42.2 for the circular, 50 for the flap, and 74.8 for those not stated.— These results not only clearly establish the importance of early amputations, but plainly show that, as regards skill in the performance of operations, and success in subsequent treatment, the Surgeons in the Confederacy can compare most favorably with those of other countries—a fact which will become all the more patent when the statistics of the war have been more thoroughly collected.— By referring to the appendix, table "C," all the facts in this regard, so industriously collected and conveniently arranged by Surgeon Sorrell, may be more accurately understood.

In the Crimea, the mortality attending amputations made in the various "thirds" of the thigh was materially different. Thus, for the *lower third* it was fifty six per cent; for the *middle*, sixty per cent; and for the *lower*, eighty six per cent. The facts which may be gathered from all of these figures, &c., are substantially as follows:

1. Amputation of the thigh is always a serious thing.

2. Primary Amputations, particularly in military Surgery, are more fortunate in their results, by far, than secondary.

3. That the danger of an unfavorable result increases for every inch as the point of Amputation, approaches the trunk, being greater for the *middle*

third, than for the *lower third*, and greater still for the *upper third*.

Taking all things into the account however, the rule in military Surgery is to operate at once, if the patient be in the Field, for such lesions as were indicated in chapter second, as justifying Amputation, since it is impossible to secure that tranquility of mind and body which is essential to the salvation of the limb. In Hospital practice, it is well to follow the advice of Baudens, and to make an effort to save the limb, if the wound be in the upper third,—where to amputate is death in a large majority of cases.

Amputation on the Hip Joint.—Directions for operating after the manner of Liston.—Administer chloroform; bring the patient's buttocks to the edge of the table; compress the antery on the *ramus* of the *pubes*; and, having inserted the long catline, at a point mid-way between the trochanter major, and the *anterior superior* spinous process of the ilium, and transfixed the limb, cut downwards and then forwards so as to form the anterior flap. Then turning the the flap back disarticulate, by severing the *capsular ligament and ligamentum-teres*, and passing the knife behind the joint, make the posterior flap by cutting downwards and backwards. This is the operation for the *left* joint. In amputating at the right joint, the knife must be entered on the inner side of the limb, just opposite the scrotum, and brought out at a point midway between the trochanter and the sup: spin: process of the ilium, while the flaps are made just as above described. Direct one of the assistants to follow the knife with his right hand, as the anterior flap is cut, so as to seize the femoral artery when divided. Ligate the the posterior arteries *first*, and then take up the femoral, with such of its branches as bleed too freely, and bring the flaps together in the usual way.

The operation is also performed by making two lateral flaps, one on the *inner* side, of the adductor muscles, and the other on the *outer* side, by putting the knife over the trochanter and cutting downwards and outwards. The inner flap is usually made *first*, and the femoral artery tied in advance—a procedure, however, which is unnecessary if the assistant be reliable. Some Surgeons prefer this operation, taking care to cut the inner flap last in order to avoid severing the artery until the outer flap has been made and the joint disarticulated.

Remarks.—This amputation can be rapidly and readily performed, though the mortality from it is very great,—so much so in fact, as almost to exclude this operation from the legitimate procedures of Surgery. When the limb is wounded high in the *upper third*, and an operation seems indispensable, it is better to make the amputation through the trochanters of the femur than at the joint. As there is great danger from hemorrhage, the flaps should not be permanently closed for some time,—not until the effects of the chloroform and the shock have entirely passed off, and reaction has occurred. Remember also to tie the ischiatic and gluteal arteries, in the posterior flap before ligating the femoral and profunda, for if the main artery is properly held by the assistant, it will not bleed. Though a single case is on record, in which the wound thus made, healed by "first intention," union does not take place even in the most favorable cases until after the most profuse and exhausting discharge of pus. Various modifications of the amputation at the hip-joint have been suggested; but the one described is incomparably the best, and it is unnecessary therefore to describe the rest.

* STATISTICS.

Reported by	Stephen Smith,	Operations	35,	Mortality	60	per ct.
" "	Henry Smith,	"	11,	"	27.3	"
" "	Legouest,	"	44,	"	90.9	"
" "	Esmarch,	"	7,	"	99	"
" "	Macleod,	"	62,	"	91.9	"
" "	Cox,	"	84,	"	75	"

This Table includes operations for disease as well as injury, a fact which reduces the per centage of mortality to considerably extent. It is agreed among military Surgeons that when performed for gunshot wounds, the mortality is over 90 per cent— thus rendering the operation so difficult as to make it the "*ultissima ratio*" of our science. In private practice when the hygienic condition of patients is good, and the proper appliances for their treatment are available, more latitude may be given in regard to the operation—though it is always an exceedingly hazardous one. All the operations of this character, performed by the English Surgeons in the Crimean war, terminated fatally, either from shock or hemorrhage, or the conjoined effects of both of them.

---

* The Author knows of but a single amputation at the hip-joint which has been performed during the present war—though he feels assured there have been others—a secondary one performed by himself, which terminated fatally. He has heard indirectly of several others, which were not more successful, but he is not prepared to speak with confidence in regard to them.

AMPUTATIONS OF THE SUPERIOR EXTREMITIES.—It must be remembered that, in injuries of the upper extremities, there is always manifested a much greater power of resistance and endurance than when the lower limbs are involved. This fact, depends mainly upon the greater vascularity of these parts, the free anastomosis of the blood vessels and the more liberal supply of nerves in the arm than in the leg, and should warn the Surgeon against the impropriety of sacrificing the one for injuries which would demand the immediate condemnation of the other. Amputation of the upper extremities is, however, frequently a surgical necessity, particularly since the introduction of conical balls, and when, from the wounding of immense numbers, the proper conservative measures cannot be adopted in order to ensure the safety of the limb.

Amputation of the Fingers.—Amputation of Fingers at the second or the last joint, should be performed precisely in accordance with the directions given for the amputation of the toes at the same points.

Amputation at the metacarpo-phalangeal articulation- These joints belong to the "Ball and Socket" variety—the phalanx furnishing the "ball" and metacarpal bone, the "socket." Turn to directions for amputating a single toe, and all the rules for this operation, will be found in detail. It is only necessary to add, that after the finger has been separated from the hand, the head of the metacarpal bone should be invariably removed, so as to give more symmetry and usefulness to the member.

The thumb may be amputated in precisely the

same manner as the fingers; but it is matter of great moment particularly to laboring men, to save it, or a portion at least, and the Surgeon should not remove it, in *its entirity*, without the gravest reasons. Its metacarpal bone may be removed for disease or injury, and the phalanges left behind, without entirely destroying its functions. When both metacarpal bone and thumb have to be removed, the operation is performed in thiswise : *Directions.* —Carry the bistoury through the soft parts between, that of the metacarpal bone and that of the fore finger until it is arrested by the trapezium above : cut the joint, by dividing the ligaments; and, then, draw the knife downwards, forming a flap of the fleshy substance which constitutes the ball of the thumb.

Amputation of the Four Fingers together.—Directions.—Pronate the hand; grasp the four fingers, placing your thumb on the joint of the little finger, and your index on that of the index finger: then make a semi-circular incision, with its convexity downwards from the inner side of the head of the fifth metacarpal bone to the outer side of the head of the second metacarpal bone. This being done, draw the knife over the four joints, so as to destroy their dorsal ligaments; divide each lateral ligament in turn; and then cut through the palmar ligaments. Lastly, slip the knife under the end of the phalanges, and cut the palmar flap, first on the side of the little finger, following exactly its palmar crease, and raising each finger in turn, to follow the knife. These are the proper directions for the *right* hand; while if the left be operated on, the knife must pass in a different direction, that is from the index to the little finger. When the operation is finished, tie the arteries, if they bleed freely; unite the wound with adhesive straps; and place the hand in a sling in a middle position.

Disarticulation of the four fingers at the Carpo Metacarpo articulation.—Directions.—Pronate the hand, and grasp the joint with your finger and thumb; make a semilunar flap on the dorsal surface of the hand, from one side to the other; divide the space between the finger and thumb in its whole length; and then divide all the dorsal ligaments transversely, except that of the second metacarpal bone, remembering not to enter the joint. All these ligaments being divided, as well as the internal and external, depress the metacarpus, and luxate the bones; finish cutting the fibrous bands which retain the joint, and also the palmar ligaments; and, then gliding the knife under the palmar surface of the bones, cut a suitable flap from it. In performing this operation, it is easy to remove the thumb, if necessary, or to retain it with either the index or little finger—which even alone are of great service.— When the operation is terminated, it only remains to tie the trunks of the radial and ulna arteries, and to bring the flaps together with adhesive straps and roller bandage.

It is a matter of the first importance that the Surgeon should have an accurate knowledge of the anatomy of the parts in performing this operation otherwise he will be compelled to use the saw in order to complete it. The terminal points of the line which corresponds with the direction of the joint, should be ascertained before commencing the opera-

tion,—the directions for which are as follows:—
Run your finger along the metacarpal bone of the
index finger until the point is reached at which it and
the second metacarpal bone approach each other,
and the former unites with the trapezoides,—this is
the inner termination of the line above referred to.
Again, trace the metacarpal bone of the little finger
upwards until a cleft is reached between the os-magnum and pisiform—a little in advance of the latter
—this is the outer terminal point sought for. The
course of this line is convex with its inclination
downwards and inwards.

Amputation at the Radio Carpal Articulation.—
The first thing to be done by the Surgeon is to
distinguish the exact seat of the joint, which may
be determined in this wise: Draw a straight line
from the point of one styloid process to the other,
and the joint will be found in the direction of a
curve, the highest point of which passes about a
quarter of an inch above the middle of the straight
line. Directions for performing the double flap operation.—Grasp the wrist so as to compress the ulna
and radial Arteries and *semi-pronate* the hand;
make a semilunar incision posteriorly, commencing
half an inch below one styloid process and terminating at the same distance below the other;—the central portion of the curve being two inches lower;
dissect up this flap and let it be drawn back by an
assistant; and, then divide the extensor and radial
tendons, the capsular ligament, the lateral ligaments, and the tendon of the carpal extensor. After this is done luxate the wrist, pass the knife behind it and cut a flap from the anterior surface, one

inch and a half long. Most Surgeons raise the handle of the knife in the last step so as to avoid including the pisiform bone in the flap, but this is unnecessary, as no inconvenience results from its being left with the skin, while the attachment of the Flex or Carpi Radialis is, in fact, preserved thereby. The radial and ulna arteries are now to be tied, if not too much retracted; the integuments closed by adhesive straps and a roller bandage lightly applied from the elbow downwards.

The *Circular Method* may be also employed, but the above process is preferable.

Amputation of the Fore arm.—Surgically, the fore arm is divisible into three portions, viz: the *inferior*, which is flattened and well suited to the flap operation; the *middle*, which is conical and favorable for the flap operation,—as it is difficult to turn the cuff of skin backward; and the *upper* third, which is round and muscular and suggestive of either the *flap*, *circular* or *oval process*.

Circular method, *Directions.*—Place the patient in a chair or upon a bed—the latter if chloroform is administered;— compress the brachial artery against the humerus by means of a tourniquet or the fingers of an assistant; and, then partly flex the fore arm and place it in a position midway between pronation and supination. The Surgeon must then place himself so as to grasp the arm above the point of amputation with his left hand, and proceed to operate according to the directions given for amputating the leg by the circular or the oval method. Apply the saw upon the face of both bones, engage the ulna first but complete its section last, as it is more firmly connected with the humerus. The ulna and radial arteries must then be tied, and sometimes the interosseous, and the wound closed with adhesive straps.

Double Flap operation.—Directions.—Place the arm in an intermediate position between *pronation* and *supination*; transfix the limb, by passing the knife either from the ulna or the radial side, in front of the bone; and then cut an anterior flap more than two inches in length, from the muscles on that side of the arm. Carry the knife to the opposite side. and transfix,—passing the instrument in at one angle of the previous wound and bringing it out at the other;—and cut downwards and backwards so as to form the posterior flap. Have the flaps raised by an assistant; cut the interosseous ligament and remaining muscular fibres; use the three tailed retractor; and saw through the bones in the manner described above. Dress as before.

The *Single Flap* operation may also be performed at any portion of the arm.

Amputation at the Elbow Joint.—The exact position of the joint may be ascertained by the following method: Find the internal condyle of the humerus,— this is *three-quarters* of an inch *above* the *articulation* of the humerus with the ulna; then, seek out the external condyle—this is about *half* an inch *above* the articulation of the humerus with the radius.— These tuberosities being on the same plane, it follows that the *articular line* is directed from *within* obliquely *outwards* and *upwards*, and that it connects two *points*, one of which is three-fourths of an inch below the internal tuberosity of the humerus, and the

AMPUTATIONS OF THE ARM.

other half an inch below the outer tuberosity. The flap, circular, and oval operations may all be performed.

The *Flap operation* as proposed by Dupuytren is undoubtably the best, even though several ligatures have to be used, and some of the first Surgeons prefer the circular method.

Directions.—Supinate and partially flex the arm; ascertain the position of the *inner* tuberosity of the humerus; and, having grasped the soft parts immediately below, pass a catline through the muscular tissue in front of the bones, entering it about one inch below the epitrochlea, and bringing it out about one-half an inch below the epicondyle; and carry the knife downwards and cut a flap at least four inches long, from the muscular tissue on the anterior face of the fore arm.

Next return to the base of the flap, and divide all the intermediate tissues by a semicircular sweep of the knife, down to the joint itself, which is entered between the ulna and radius. Then divide all the ligaments; dislocate the joint; and either cut off the olecranon process with a saw, or pass the knife behind it and remove it with the rest of the bone.

Amputation of the upper Arm.—The circular, oval, double flap, or single flap operation may be made, though the circular and double flap are most popular.

Directions for the Circular Operation.—Raise the arm almost at a right angle; divide the skin by a circular incision; retract forcibly and divide the superficial fibres of the muscles; retract again forcibly and divide the deeper fibres down to the bone; denude the bone for a short distance upwards; and, then, having retracted the soft parts sufficiently, saw through the bone. Or, again the integuments may be dissected up, and turned back in the form of a cuff, and the muscular fibres divided, at one sweep down to the bone as was previously described in connexion with the general considerations of the circular mode of amputation.

Tie the brachial artery, which is found at the inner margin of the biceps, and such of its branches as may bleed too freely; then bring the wound together with adhesive straps, and unite the integuments in an oblique direction.

The double flap operation: Directions,—Arrange all the preliminaries as for the previous operation; seize the limb with the left hand; transfix it anteriorly; and cut a flap at least three inches in length. Carry the knife behind, and pass it through the arm at the upper angles of the previous wound; and cut a flap slightly longer than the first; pull the flaps back; divide all the tissues to the bone; retract and saw through the humerus. Tie the brachial artery and its branches, and bring the flaps together with adhesive straps and sutures.

Amputation at the Shoulder Joint.—There are several different procedures recommended in this connexion, among which that of Larrey is incomparably the best, though the method of Lisfranc has many admirers. Both of these methods will be described in detail, and the Surgeon left to select that one which he deems most likely to meet the presenting indications.

Directions for performing Larrey's Operation.—

Compress the subclavian artery in its *outer* portion, just above the clavicle, by means of a key; and then make an incision from the border of the acromion, to one inch below the level of the neck of the humerus—dividing the integuments and separating the deltoid into two equal parts. This being done, make two oblique cuts, from the first incision on either side, and terminating, the one at the anterior border of the axilla, and the other, at its posterior border, and both prolonged in such a manner as to divide the tendons of the pectoralis major, and the latissimus dorsi near their insertions; divide the tissue which retains these two flaps, down to the bone; and draw them back so as to expose the joint; cut through the capsule and tendons above, and on either side; dislocate the head of the bone *outwards*, by carrying the arm transversely across the body, either forwards or backwards; pass the knife behind it, so as to separate it completely from the soft parts; and finish the operation by cutting the skin and soft parts transversely, and on a level with the inferior edges of the oblique incisions. Tie the arteries, beginning with the axillary, and bring the edges of the wound together with straps and sutures. This description is long, but the operation itself may be executed with great celerity and neatness.

Directions for performing Lisfranc's operation,—slightly modified.—Compress the artery and place the patient according to previous direction; lay hold of the arm a little above the elbow and move it from the side and slightly backwards so as to give a view of the skin in the Axilla; then push a long sharp pointed and narrow knife, through the skin in the arm pit, immediately in front of the tendons of the latissimus, dorsi and teres major muscles, and bring it out a little in front of the extremity of the acromion process—taking care to move the elbow outwards, upwards and backwards, as the thrust is made; and with the arm in this attitude, carry the knife with a sawing motion, downwards, backwards and outwards, so as to form a flap at least four inches long, of the posterior portion of the deltoid, of the tendons of the lattissimus, and teres, and of the skin. Raise the flap; divide the heads of the muscles surrounding the joint; carry the elbow in front of the chest and cut through the capsular ligament; disarticulate; and then, with the knife passed in front of the bone, form another flap by dividing the muscles and integuments. The axillary artery is thus divided, and, to prevent hemorrhage, an assistant should grasp the soft parts of the axilla at this stage of the operation.

As soon as the limb is detached, ligate the main vessel, together with the circumflex, subscapular, and such other arteries as may bleed freely; then bring the edges in apposition and confine them with the usual appliances, taking care to have the line of union perpendicular.

The operation just described, is for the *left* limb, and it is necessary to modify it for the *right* by making the first flap from above, downwards and backwards, and, then, continuing, as directed above.

Remarks.—In operations upon the hand, it is important to preserve as many fingers as possible, and particularly the thumb. If the head of the metacarpal bone be not removed, when a finger is amputa-

apparent in connexion with this operation. Thus, Guthrie reports 19 cases of secondary amputations at the shoulder with a mortality of 19, and 19 cases of primary amputations with only 1 death. Dr. Thompson, in giving his experience after the Battle of Waterloo, states that almost all of those recovered who had undergone primary amputation at the shoulder joint, while fully one half died of those on whom it became necessary to operate at a later period. Legoust, Gualla, Smith, Esmarch and Macleod all agree, that whereas, the mortality attendant upon primary amputations at the shoulder joint was not more than 33 per cent, the mortality following secondary amputation was at least 75 per cent. This operation admits of no delay, and if performed at all must be done quickly in order to give the patient a chance for his life.

The experience of modern Surgery has demonstrated the fact that resections at the shoulder joint are not only safer than amputations, but may take the place of them in a large majority of cases,—thus preventing great deformity, and securing a comparatively useful member. This subject will be more freely considered under another head, and for the present, it is sufficient to say, that such operations have been attended with wonderful success, according to the testimony of Percy, Baudens, Legoust, Esmarch, and Macleod, even when several inches of the shaft of the humerus had been destroyed.

Amputations at the shoulder joint have been frequently performed by Confederate Surgeons during the present war, and with decided success, though sufficient statistical information has not yet been gathered to determine the relative per centage of mortality which has attended the operations in their hands.

The Shock attending amputation at this articulation is great, and should always be provided against by a liberal allowance of brandy or whiskey, before and during the administration of the chloroform.

It is earnestly to be hoped that the Surgeons of our army wil not content themselves simply with the successful performance of amputations, but that they will keep accurate records of their cases, noting whatever is of importance connected with them, establishing the relative per centage of mortality for the different operations, and contributing, of their wide and varied experience, something at least towards the advancement and perfection of the science of military surgery.

For the number of operations performed in Richmond after the battles of the "Seven Pines" and the "Seven Days" the reader is referred to Table " F " of the appendix to this volume.

ted at the metacarpo-phalangeal joint, there will always be deformity, whereas, if this precaution is observed, the symmetry and usefulness of the hand can, in a great measure, be preserved. Sorrell reports two successful disarticulations of the wrist joint.

In amputating the fore-arm every effort should be made to preserve as much of the member as possible with a view to its future usefulness. Great care should be taken to have the bones of equal length and exactly parallel, for otherwise they will protrude through the flaps, producing ulceration, or, it may be, a conical stump. From the 1st of April to the end of the Crimean war, Macleod reports 54 operations, with only *three* deaths; Dr. Lente reports 39 operations with four deaths, Dr. Haywood reports 6 with one death, and Sorrell records 45, with only 6 deaths, and 39 recoveries—all going to show that the rate of mortality attending it is very low. For farther particulars, consult Table " D " of the Appendix.

All other things being equal, it would be better to amputate just below the elbow-joint rather than through it; but, the danger to the patient, from subsequent inflammation which is likely to involve the articulation, should always be taken into the account by the Surgeon. Amputation at the elbow-joint was first performed by Ambrose Paré but subsequently sank into disrepute. It has since been revived by Dupuytren and Velpeau, and may be performed with propriety when the operator desires to avoid the danger from inflammation referred to above, or to preserve a more useful member than an operation above that point would allow. The operator should never forget that the articulation proper is below the tuberosities of the humerus, and that he may be readily misled by appearances into tranfixing too high and thus making the flaps too short to cover the head of the bone.

Amputation of the upper arm can be very readily performed and is attended with great success-

STATISTICS.

Reported by Macleod, Operations 102, Mortality 24.5 per cent.
"   "   Norris,   "   32,   "   6.3   "
"   "   Lente,   "   58,   "   40   "
"   "   Sorrell,   "   192,   "   28   "
"   "   Haywood,   "   4,   "   00   "

The reader is referred to table "E" of the appendix.

To Barron Larrey belongs the credit of having elevated amputation at the *shoulder joint* to its proper rank in the art of Surgery; and the subsequent experience particularly of military Surgeons has demonstrated the correctness of his views in regard to it.

STATISTICS.

Reported by Macleod, no. of operations 59, no. of deaths 13.
"   "   Buel,   "   39,   "   13.
"   "   Lente,   "   19,   "   11.
"   "   Gross,   "   25,   "   12.

As this operation is generally performed in connexion with some wound of the body of greater or less magnitude, whereby its result is materially controlled, statistical tables cannot afford a just estimate of its intrinsic value. This fact should be borne in mind by the military Surgeon particularly, as the question of the propriety and results of this operation must constantly present itself in field service.

The value of primary amputation is particularly

same plane with the external, a smaller instrument should be introduced to cut through it, in order to avoid tearing the dura-mater at one point before another.

7. Should the sinuses be opened, hemorrhage can be arrested by plugging.

8. Should the *middle meningeal* artery be divided, the hemorrhage is serious, and difficult to control. Compress it with a bit of lint, placed inside the cranium, and retained by a thread, or with a plate of lead bent so as to embrace both surfaces of the bone; or by plugging it with sealing wax. Larrey touched the bleeding orifice with a steel probe heated to whiteness; while Dorsey and others recommended the application of a ligature.

Operation on the bones of the Cranium.—This may be divided into five different parts, viz: denudation of the bone; perforation of the bone; removal of the detached piece of bone; removal of the cause of compression; dressing and after treatment.

Denudation of the bone. — Directions.—The point of the cranium upon which the operation is to be performed, having been shaved of its hair, and the patient placed in proper position, divide the soft parts by means of a crucial or semilunar incision; dissect up the flaps, revert them, and have them held out of the way by an assistant, and control the hemorrhage from the severed vessels, either by applications of cold water, twisting them, or ligatures.

The first incision should reach to the bone, a

the flaps wrapped in fine linen to prevent injury to them.

Perforation of the bone.—This is to be accomplished either with the Hand Trephine, or the Trepan instrument of Hildanus, which may be made to revolve by a brace, or like a drill by means of a bow. The former is preferable, as the Surgeon can control it better.‡ *Directions.*—Introduce the *pyramid* or *central bit* beyond the level of the crown of the instrument, firmly secure it by means of the screw attached to the side for that purpose, and enter the trephine into the bone with a semicircular motion of the hand, until the teeth of the saw have reached the external table and made for themselves a furrow in it. Now retract the pyramid, lest it injure the dura-mater; continue the rotary motion, holding the instrument perpendicularly to the bone, withdrawing from time to time, to clean its teeth with the brush and to enable the Surgeon to sound the depth of the groove;—and penetrate both the diploe and the internal table. When the instrument has penetrated at several points, introduce an elevator into

---

‡Dr. G. A. D. Galt of the Confederate Army has devised a new Trephine, with the object of avoiding injury to the membranes and substance of the brain. The instrument consists of a truncated cone with peripheral teeth arranged in a spiral direction, and oblique crown teeth. When applied the peripheral teeth act as a cutting wedge so long as the counteracting pressure acts on the crown teeth. On the removal of the pressure by the division of the Cranial Walls, its tendency is to act on the principle of a screw, but, owing to its conical form and the direction of its peripheral teeth, the action ceases and the instrument penetrates no farther. Dr. Galt says that he has operated on the dead subject twenty times, and has never succeeded in wounding the membrane, although he has endeavored to do so. Subsequent practical experience has demonstrated the great utility of the instrument.

CHAPTER IV.

EXCISION OF BONES AND JOINTS.

The instruments required in this connexion are the Saw—Hey's, Chain and Circular;—cutting forceps; perforator; mallet; chisel; gouge; rasp; elevator; scalpel, &c.

By the circular saw is meant the trephine, an instrument potent for good or evil, according to the necessities of the case and the skill and judgment of the Surgeon.

Trephining.—The circumstances under which this operation has been recommended, are the following: fracture of the skull with depression of the bone; fracture of the bone with penetrating wound of the dura matter; epilepsy, depending upon depression, or upon the existence of some point of irritation of the skull; and the presence of foreign bodies in the cerebral substance, including effused blood and pus.

Locality of the Operation.—Avoid the sutures; those parts of the skull immediately over important arteries and veins; those regions of the skull where the two tables are situated at some distance from each other; the thicker portions of the bone; and the part immediately under the temporal muscle.

The occipital protuberance, meningeal artery, and the sinuses are particularly to be avoided.

Mode of Applying the Instrument.—In simple fracture apply the instrument with the pyramid resting near one margin of the fissure so the section may extend on both sides. In fractures with depression, see that the crown of the instrument does not rest upon the loosened bone, for fear of causing laceration or irritation of the soft parts beneath.

When a foreign body is wedged in a wound and the fracture is but limited, the crown of the trephine should embrace the whole solution of continuity.

In the case of extravasated fluids, operate immediately over the seat of effusion, which is frequently on the opposite side from the wound.

Position of the Patient.—Make the patient assume a recumbent position, with his head resting upon a well cushioned board, and firmly held by an assistant.

General Rules.—1. Do not operate simply for the injury, but for the consequences produced by it.

2. Do not be hasty in resorting to the operation, but wait for nature and time to do their work.

3 If the operation be not performed before the development of inflammatory reaction, wait for its subsidence.

4. Bear in mind that in the young the skull is more yielding than in the old, and more readily depressed without fracture.

5. In caries and necrosis it is deemed most prudent to permit the diseased portions to separate themselves until they can be seized with the forceps and extracted.

6. When it becomes necessary to trepan the frontal sinus, the internal table not being on the

the groove, and seperate the circular piece from the internal table.

The division of the diploe can be readily recognized by the ease with which the instrument penetrates its substance and the bloody detritus which escapes. This structure is deficient in children and old persons,—a fact which should be remembered in operating upon them.

When the trephine has to be applied so as to cover a small fractured portion of the skull, or some foreign body lodged in the bone, the perforator can be used to start the crown; and a piece of wood, cork or sole leather with a hole in it of the proper size, and firmly held by an assistant, will serve to retain the instrument in position until the teeth have made a sufficient groove in the bone.

Where fractures exists with depression, and the margin of one bone overlaps the other; where there is depression without fracture, and where an enlargement of the angular fissue has to be effected, an opening may be made with Hey's saw. A piece of leather with a crevice cut into it, must be placed on the skull, within which the straight end of the saw plays until it sufficiently introduces itself.

Removal of the detached piece of bone.—Fasten the bone screw into the orifice made by the central pin, and by a few lateral motions detach the piece. It is better to introduce the elevators on the opposite sides of the piece so as to separate and lift it out. Sometimes it is brought out with the trephine itself. If prominent points of bone remain, they should be carefully removed with a lenticular knife or Heys' saw.

To remove the cause of Compression—If it be desirable to raise a portion of the bone,—as for fracture with depression—, introduce the Common Elevator between the cranium and dura mater, without dividing the membrane, and gradually elevate the depressed portion by using the opposite margin of the bone, or the finger as a fulcrum. Loose portions of the bone are to be picked away with the forceps and if the operation has been undertaken for the removal of a ball or any foreign substance, it may be seized with the forceps and drawn out, unless too much effort be required to bring it away.

If the operation be early done for extravasition or effusion, the fluid, if on the outer side of the dura-mater, will come away of itself. Should it prove to be blood however, it must be broken up with the finger and then removed. If the fluid be below the dura-mater, this membrane will be found detached from the bone, and of a brownish hue with a *bulging* at some particular point and a feeling of fluctuation below. To remove this fluid the dura-mater should be punctured, by pushing a straight sharp pointed bistoury through it.

Dupuytren plunged his knife deep into the ceretral substance itself, and opened an abscess more than an inch from the surface. His example has been followed by other eminent Surgeons, but it is too bold and dangerous a measure for universal imitation.

The Dressing and after Treatment.—Apply cold water dressings, instead of the cerate &c., recommended by older Surgeons. Do not disturb the wound until suppuration ensues, when it may be washed and carefully dressed twice daily.

Remarks.—As late as the eighteenth century Trephining was practised in almost every variety of wounds of the head, both as a curative and a preventive measure—or a means of protection before dangerous symptoms were developed. The Trephine was used on all occasions and for every possible injury, realizing, even as far as the most eminent Surgeons were concerned, the lines to Sidrophel,—

> "He used trephining of the skull,
> As often as the moon was full."

This shameful misapplication has been most energetically and successfully opposed by Desault, Abernethy, Langenbeck, Physick, Cooper and others until more rational, as well as safer views, are entertained in regard to the operation by the whole Profession. The reaction against the use of the instrument upon the cranial bones has gone so far that some have rejected it altogether as dangerous and unnecessary under all circumstances; but Sir A. Cooper, and Sir B. Brodie have very clearly demonstrated the impropriety of this conclusion so far at least as some cases of compound fractures with depression are concerned.

In military Surgery, the trephine is far less used than formerly,—and the experience of Stromyer Macleod, Hewitt, Guthrie, Cole and Chisolm clearly demonstrates not only its inutility in the treatment of cranial wounds generally, but the positive detriment resulting from its employment even in many cases of fracture with depression and compound fracture, for which it has heretofore been primarily recommended. Stromyer, who was Surgeon in chief in the Schliswig—Holstein Army, and "one of the highest authorities in Gun-shot wounds of the head," positively and peremptorily affirms, "*that in military Surgery, trephining is never needed.*" This opinion is endorsed by Loeffler, and in a great measure sustained by Chisolm, and other more modern military Surgeons.

It is now well known that a depression of the *outer* table does not necessarily indicate a corresponding depression of the *inner* tablet, and that both tablets may be so depressed as materially to compress the brain without interfering with the functions of that organ, or developing an unfavorable symptom. Trephining is also known to be a serious operation—to be nothing "more or less than boring a hole in a man's skull"—and as calculated, even under favorable circumstances, to produce irritation and inflammation of the delicate membrane it exposes, and of the sensitive cerebral substance beneath. These facts, taken in connexion with the recorded experience of the great authorities previously referred to, should teach the military Surgeon. the vast importance of deliberating well before resorting to this operation, and of only employing it when all other means have failed to produce those results upon which the salvation of his patient's life depends. He should avoid all *haste* in its employment, waiting for nature, assisted by other more rational and less violent remedies, to relieve the symptoms of cerebral compression, and to restore the patient to his normal condition. If, however, his expectations in this regard are disappointed,—if sensibility and motion fail to return, while coma and stertor increase, in despite of the most energetic antiphlogistic measures, showing such an augmentation of congestion in the cerebral substance as immediately jeopardizes the patient's life, then, the Surgeon *may* resort to the trephine as a " forlorn hope" whether the fracture be simple, compound, or comminuted, or whatever *the* nature and limits of the injury. He should neither endanger his patient's life by resorting too hurriedly or indiscreetly to the Instrument, nor permit him to die for the want of it through an unbecoming timidity, or a slavish subserviency to fashion and authority.

The trephine may also be applied successfully to any one of the long bones, when attacked with caries or necrosis, and for the purpose of removing foreign bodies impacted in them, such as balls, pieces of metal, &c.

The operation is nearly the same in these cases as that just described, only differing according to the depth, density and form of the the bone.

The experiments and observations of modern military Surgeons are decidedly favorable to resection, particularly in those cases where the choice is between the removal of a joint and amputation above it. Primary resections have been found **equally as important as primary amputations.**

RESECTIONS IN GENERAL.—Resections are undertaken
1. For the removal of the articulations alone.
2. For the removal of the shafts of bones.
2. For the exterpation of certain bones entire.

The circumstances which justify resection are:

1. Caries of the articular extremeties when other means have failed.
2. Osteo-sarcoma, spina-ventosa, and malignant affections generally.
3. Compound and comminuted fractures, such particularly as are caused by conical balls, impinging either upon the shaft or articular surfaces of bones. Also the protrusion of fragments through the skin, when they cannot be replaced, or are denuded of their periosteum.
4. Compound luxations, when insurmountable obstacles present themselves to reduction.
5. Necrosis of bone, when elimination is tardy.
6. Projection of the end of a bone beyond the stump in badly performed operations.
7. Exostosis, or when some foreign body has lodged in the bone and cannot be removed.

Resection should never be attempted unless the patient has manifestly strength enough to bear a difficult operation and a tardy convalescence, and it is therefore, contraindicated when there are symptoms present of any one of the cachexiæ; of unusual nervous susceptibility, or of marasmus. It is also frequently exceedingly difficult, in chronic affections of the joints to distinguish between vessels, nerves, &c.; and hence there is danger of tetanus, protracted suppuration, purulent absorption, and erysipelas.

Rules for Resections in general.—Distinguish well the anatomical relations of the parts before commencing the operation. Know where nerves, and vessels are to be found, for it is exceedingly difficult to distinguish them during the resection.

2. In addition to the ordinary instruments, have on hand, a cutting forceps, a gouge, a mallet, and saws of different sizes and shapes.

3. Open a free way to the bone, but expose as little as possible of the muscles and tendons.

4. The nerves, the veins and the arterial trunks are never to be divided; while the tendons, as a general thing, must be preserved.

5. Before employing the saw, ascertain to what extent the bone is diseased, and see that the soft parts are well out of the way of injury.

6. Remove completely every part touched by the disease or reached by the injury.

7. Cut off the bones connected with the articulations at the same distance from the joint.

8. Preserve as much of the periosteum and take away as large a portion of synovial membrane as practicable.

9. When a lower limb has been operated upon, bring the bones together, and extend it; but when an upper, put it in a state of semiflexion, and leave the bones a little apart so as to secure, if possible an artificial joint.

10. Make the incisions on the side opposite to the main arteries.

11. Make the existing wound lie, if possible, in the line of one of the incisions, which should be so arranged as to permit the free drainage of pus.

PARTICULAR RESECTIONS.—*Resection of the bones of the upper limb.*

Resection of the Metacarpo-phalangeal articulation.—Either the head of the metacarpal bone or the end of the phalanx may be removed. Directions; commence half an inch from the point at which the saw is to be applied, and make a flap with its base towards the finger; dissect up this flap; turn aside the extensor tendon and separate the muscles from the bone: open the joint carefully, so as not to divide the flexor tendons; disarticulate and isolate the diseased portion; and then slip a small peice of wood or a spatula under the bone, and saw it off.

Extraction of the First Metacarpal Bone.—Directions.—Make an incision along its radial border, extending half an inch beyond each articulation; cautiously detach the skin and tendon from its dorsal surface and the muscles from its palmar face: have the edges held well apart and carry the knife through the upper articulation: then luxate the bone outwards and pass the knife completely along its inner surface; and finally carry the knife through its lower articulation. The radial artery may be avoided, but if cut it can be readily ligated. Close the wound, and keep the parts in their normal position.

The other bones of the metacarpus may be removed by following the same general plan.

Resection of the Wrist Joint.—Directions.—Make two longitudinal incisions, terminating on a level with the articulation, one along the outer side of the radius, and the other along the innerside of the ulna, near their anterior edges; unite them by a transverse incision across the back of the wrist; dissect up this quadrilateral flap, avoiding the tendons which glide in the grooves of the bone; draw the tendons aside, as much as possible, and detach the soft parts; and then pass a spatula under the bones and saw both ulna and radius at once. Bring the parts together and treat on general principles. The tendons which control the motins of the joint may be divided in an emergency and the knife passed more directly into the joint, as it is not expected to preserve the movements of the articulation after the operation.

Remarks.—In consequence of the close connexion of this articulation with the flexor and extensor tendons, consolidation of these and their sheaths is likely to occur, together with the consequent loss of motion in the hand. Many cases, however, will be found in which, even with a stiff wrist, there may be some motion of the fingers; and with all the disadvantages attending this operation, it is far better to have a hand, whatever may be its condition, as regards mobility, than no hand at all. The lower extremities of the radius and ulna may be excised; while the carpus remains intact, if these bones alone are involved in the the disease or injury, by simply following the first steps of the operation, as above pointed out.

Extirpation of the Radius.—Directions.—Semiflex the arm; make a longitudinal incision on the external anterior border of the radius, so as to lay it bare; dissect back the integuments; push the soft parts aside; pass a director or scalpel under the bone and saw through it; clear the fragments from the soft parts; and then separate them from their articulations, avoiding the nerves and arteries.

Resection of the Body of the Ulna.—Directions.—
Make a transverse incision down to the bone, four inches and a half below the olecranon, and extending a little more than half the diameter of the arm; make another longitudinal incision, intersecting the lower part of the former, and along the most superficial portion of the bone down to the wrist joint; commence at the first incision and dissect the soft parts around the bone for three inches; insert a spatula and saw through the bone transversely; continue the dissection to the wrist joint; and then disarticulate and remove the bone. Avoid wounding the ulna nerve, and tie the ulna and interosseous arteries if divided.

The inferior extremity of the ulna may be resected by making a longitudinal incision along the border of the ulna; then making another longitudinal incision across the back of the joint; dissecting up the flap and turning it back; drawing aside the tendons; and disarticulating.

Remarks—Several cases are recorded of successful removal of the ulna and radius, and the results attending the operation are such as to warrant the Surgeon in resorting to it under some circumstances. It may be more advantageously attempted for disease than for injury, as the soft parts are less likely to be involved in the first instance than in the last. Resection of the radius is more likely to interfere with the mobility and symmetry of the arm, than the removal of the ulna, for obvious reasons.

RESECTION OF THE ELBOW JOINT.—Surgeons are much divided as to the best operation for the excision of this joint, some advocating the ⊢, and others the H shaped incision; while still another class prefer Bucks modification of the latter, which consists of two longitudinal incisions, the horizontal cut being omitted and the sides directed so as to expose the bone without dividing the attachment of the tendon of the triceps. Ordinarily the following plan will be found the most available. Directions: Place the Patient on his face, near a well lighted window, upon a table four feet high, so that his arm is supported and presents to the Surgeon the posterior and internal face of the articulation; then, make an H shaped incision, taking in the breadth of the articulation, exposing the heads of the bones, and dividing the skin and tendon of the triceps; dissect back these flaps carefully, taking care to remove the ulna nerve from its bed on the inner side of the arm, behind the epitrochlea; divide the posterior ligaments and expose the joint, separate the soft parts carefully, avoiding the nerves and arteries; pass the handle of a scalpel under the humerus; saw off and detac as large a portion of the humerus as may be necessary; and finally, attack the bones of the forearm and remove such portions of these as may be necessary,—remembering, that if either one of them is not implicated in the disease or injury, to leave it unmolested.

Then close the lips of the wound by means of

sutures; leave the bones slightly separated; keep the limb upon pillows and rely exclusively on the cold water dressing. Remember, however, to prevent anchylosis by passive motion of the joint when the soft parts have cicatrized. It is a matter of great consequence not to remove more of the bones than is absolutely necessary.

The shaft of the humerus should not be encroached upon, if it is possible to avoid doing so, or the excision of the radius and ulna carried below the insertion of the brachialis anticus and triceps. The position of the parts, and the relations of the bones to each other should be scrupulously attended to, bagging of matter prevented and exuberant granulations repressed.

Remarks.—This is comparatively a modern procedure, having been suggested by Park of England in 1781, and performed by Moreau in 1782. It is to Roux, Crampton and Syme, however, that the profession is indebted for the revival and vindication of this operation—thus securing moveable joints and comparatively useful members to many, who otherwise would have been deprived of their arms. As a proof of the utility of this operation, it is only necessary to refer to a few facts, which have been collected in regard to it. During the Schleswing-Holstein campaign, Langenbeck and Stromyer, report that of fifty four amputations of the arm, nineteen died, whereas of *forty* resections, performed under identical circumstances, and with similar appliances for operating, dressing and transporting, only *six* died. Macleod records 20 operations for excision of this joint, and *seven* deaths,—

four following secondary resections, and not being connected with the operation. To this might be added the testimony of hundreds in civil practice, both in America and Europe; whilst if the experience of Confederate Surgeons were properly collected, the value of this operation would be rendered still more apparent to the medical world.

The importance of Primary resections is particularly conspicuous. Thus, of *eleven* cases excised within twenty four hours before reaction had set in, but *one* died; of *twenty* cases between the second and fourth day, or during the stage of irritation and excitement, four died; and of *nine* cases operated upon between the eighth and thirty seventh day, only *one* died.

The necessity for this operation is not so great when the joint has been opened by a sabre cut, as when a ball has passed through it, grinding up the bones, annihilating the ligaments, and completely destroying the articulation. Sorrell reports six operations and four deaths. Wounds of this joint may be readily recognized by the following circumstances: the facility with which the interior of the joint can be reached by the probe or finger; the general direction of the wound; preter natural mobility or entire loss of motion; and the escape of synovia—circumstances which should always be taken into the account because of the imperative necessity for promptness in the performance of the operation if the wound has really involved the articulation.

Resection of the Shoulder Joint.—The operations of White, Lisfranc, and Syme have all their advocates, but the following plan, is perhaps the best as a general rule.

Directions.—Compress the Subclavian Artery above the clavicle; make a V shaped Flap of the deltoid muscle about three inches long, beginning at the acromion process and terminating on the upper and outer portion of the arm; dissect this up and expose the capsular ligament of the joint; ligate the circumflex arteries which are divided in the first incision; carry the arm over the chest; divide the capsular ligament and turn the head of the humerus out of the glenoid cavity; remove the long head of the biceps from its groove; place a spatula behind the bone; and then remove with the injured portion of the humerus.

This being done, return the flap to its proper position; place the patient in bed; support the arm upon soft pillows; and apply cold water dressings. The most tedious, and perhaps embarrassing portion of the operation, is the removal of the tendon of the biceps from its bed. Rather than prolong the sufferings and dangers of the patient unnecessarily, it is better to sever this tendon, and to conclude the operation, as experience demonstrates, that no serious inconvenience results from such a course. The deltoid is usually paralyzed after this resection even when White's operation is performed but the other muscles surrounding the joint form new relations, and a very useful limb is secured to the patient—though considerably shortened, and somewhat deformed. It is best to place the limb upon a pillow or a long, broad splint, without applying bandages, and to keep the patient perfectly quiet until the inflammatory stage has past, and suppuration has been established,—when, with his arm carefully placed in a sling, he may be permitted to walk about. If suppuration be excessive, sustain his strength by the free use of stimulants and a liberal diet.

Remarks.—This operation is usually successful, more so perhaps than most of the Amputations, and, as it can be readily performed, it should commend itself particularly to the attention of military Surgeons.

STATISTICS.

Reported by	Larrey,	Operations	10	deaths	4*
"	Baudens,	"	14	"	1
"	Stromyer,	"	18	"	7
"	Legouest,	"	6	"	4
"	Macleod,	"	14	"	1
"	Sorrell,	"	5	"	2

Rules to be observed in Resection of the shoulder Joint:

1. Perform primary rather than secondary operations. It must not be forgotten, however, that secondary resections of this joint though inferior to primary, are, according to Stomeyer and Esmarch, more successful than those of other joints.

2. If upon a proper examination only a portion of the head of the humerus is found injured, remove that and leave the remainder intact. This will at least facilitate the healing of the wound, if it does not secure so useful a member.

3. The whole head and a considerable portion

---

*None were directly attributable to the operation. Thus 2 died of Scorbutus, 1 of Hospital Fever, and 1 of Pest after recovery.

of the shaft may be removed, with advantage, if implicated. Thus, though Guthrie believed that the insertion of the deltoid was the lowest point at which the bone should be divided, Esmarch has shown that at least four and a half inches can be removed and yet a most useful arm remain.

4. The U incision facilitates the performance of the operation, but the straight incision of White secures a greater degree of motion, as it does less injury to the deltoid muscle. It is important, however, in gun-shot wounds to include the two openings in the incision.

5. Arrange the line of incision so as to give free exit to the pus which is produced in large quantities, so as to avoid sinuses and abscesses in the neigborhood of the joint.

6. In field practice it is not necessary to make the incision so extensive, as under other circumstances. The muscles and tendons being thus preserved afford a better chance of restoring the action of the limb, while the healing process goes on with more rapidity and success.

7. If the head of the humerus be entirely detached, and thereby increase the difficulty of disarticulation, it may be seized with the fingers, or a pair of strong bullet forceps, and the manipulation facilitated.

8. It is well to compress the artery above the clavicle, or to have arrangements made to do so with celerity and success in the event of too great a flow of blood.

9. Avoid wounding the nerves, vessels and the glenoid cavity during the operation.

10. Never operate on the field proper, unless there are facilities on hand for supporting the limb and for transporting the patient to some neighboring hospital.

11. Support the patient's system, both as a means of relieving or preventing shock, and of securing that "plasticité" of constitution upon which a speedy convalescence and a proper union so much depend.

Stomeyer prefers a semicircular incision over the posterior surface of the articulation; Langenbeck favors one straight incision on the anterior aspect of the joint; Franke and Schleswic add to this a transverse cut; Baudens makes a straight incision on the inside of the arm; Macleod inclines to the perpendicular cut of White immediately through the deltoid; while Chisolm advocates the U shaped incision described in the preceding pages of the work.

The dangers of cutting across the fibres of the deltoid, are for the most part imaginary, in as much as, according to the practical experience of all who have witnessed and practised this operation, the fibres of that muscle speedily form unions which give them control over the arm to a very considerable extent.

Velpeau seems to have been particularly unfortunate with this operation. He reports thirteen deaths from it, and says that many more have occurred within his knowledge. The weight of testimony is, however, decidedly in favor of it; and when the experience of Confederate Surgeons is accurately recorded the weight of testimony in its favor will be overwhelming.

Resections of the Clavicle and Scapula.—Circumstances occassionally demand these operations, though they are of rare occurrance. Watt has resected the entire clavicle by making three incisions, circumscribing a quadrilateral flap, and disarticulating the bone at either extremity. The Scapula has also been removed in its totality, but it is too difficult and dangerous an operation to be repeated save in the most extraordinary cases.

There are no general rules for these operations, but each Surgeon, relying upon his knowledge of anatomy and his acquaintance with surgical principles generally, must proceed as his judgment dictates.

RESECTION OF THE BONES OF THE LOWER EXTREMITY.—Resections of these bones are not so successful as of the upper extremity.

Resection of the Anterior End of the first *Metatarsal bone.*—Directions: Cut a flap on the inside with its base posterior; denude the bone to the joint at which it is to be cut; saw it *perpendicularly* to its axis; detach it from the soft parts; and then separate it from the phalanx. This bone has been completely exterpated by Malgaigne.

Excision and Resection of the Bones of the Tarsus.— No precise rules can be established, but the operation is easy and the result satisfactory. The space left by the removal of the bone is filled up by matter which subsequently ossifies, and thus, in a measure, prevents deformity and enables the patient eventually, to walk well. Caries, or necrosis of the oscalcis is a serious circumstance, since, when its inferior surface is excised, the equilibrium of the body is destroyed, and the weight thrown forward on the point of the foot; while if the tendo, Achillis is cut, great inconvenience results. But even with these disadvantages, resection is better than amputation, as the limb is saved, and the patient can walk, however, imperfectly.

The astragalus may also be extirpated, by luxating the bone through the integuments, and dividing its attachments; but the state of the parts must furnish the proper guide to the Surgeon. After this operation the foot is fixed to the Leg and the resulting lameness great.

Resection of the Ankle Joint.—Directions.—Make an incision three inches long, from the inferior and posterior portion of the outer malleolus; from the lower end of this cut, make another transversely forwards and only dividing the skin; dissect back the flap and disengage the fibula; and seperate the external malleolus from the other bones with the chissel and mallet,—not employing the saw, because as there is no interoseous space, nothing can be introduced behind the bones so as to protect the soft parts. Dress in the usual way.

Remarks.—Resection of this joint has not succeeded so well as that of the others mentioned above, or even of the knee. It is recommended by the teachings of conservative Surgery; but the experience of the profession is against its practical utility, and amputation is now regarded as decidedly preferable.

Removal of the Fibula.—This bone may be removed either in its entirety, or partially.

Directions.—Make an incision three inches long on the inferior portion of the bone, or for its whole length; detach the soft parts as high up as the operation is to be performed; divide the bone; and, then detach it from its articulation with the tibia,—taking care to cut as close as possible to the bone so as to avoid the anterior tibial artery. The same general plan may be followed for the removal of the upper portion of the bone, or for the whole of it. Portions of the tibia may be removed on the same general plan.

Resection of the Knee Joint.—This operation was first performed by Park in 1781, and has been variously modified by Moreau, Begin, and Syme.

Among the various processes proposed in this connexion, the following offers the most decided advantages.

Directions.—Bind the leg at a right angle to the thigh; make a transverse incision slightly curved and with its convexity downwards, under the patella cutting into the articulation; make then, two longitudinal incisions upon either side of the limb and perpendicular to the first; dissect up this flap, including in it the patella; destroy first the lateral and the posterior ligaments; carefully detach the soft parts from the femur then pass a a wooden splint or piece of thick leather under it; and remove the injured or diseased portion with the saw. If the *tibia* be injured or affected, extend the perpendicular flap; separate the soft parts from the tibia, and remove a portion of it with the saw.

The patella may be removed or not according to the judgment of the Surgeon. Syme advises its removal, but on the other hand, Pancoast declares that it should unquestionably be left, "as it will serve to furnish a broader basis for the subsequent union of the bones."

If any small arteries are cut they should be immediately ligated;—the parts brought carefully together; and a hollowed splint then applied to the posterior surface of the limb, extending from the buttocks to the heel,—while cold water dressings are applied to the wound. It is useless to expect a speedy convalescence; and it is not improbable, that profuse suppuration, numerous abscesses, and exfoliation of bone may present themselves at some period in the history of the case; but if the Surgeon will watch the patient closely and see that his system is kept up to its normal tone, &c., a favorable result may eventually be predicted and obtained in some cases.

Remarks.—This operation is a modern one, dating back only to the year 1781, when it was first performed by Park, and has not yet received the endorsement of the profession. A few remarks as to the relative value of resections of the knee joint, will not be out of place in this connexion. It is well known that, when this joint is opened, whatever the extent of the injury or the nature of the missile inflicting it, violent inflammation of the synovial membrane lining the articular cavity and of the tissues surrounding its exterior, speedily follows, accompanied by great pain, excessive heat, considerable tumefaction, and violent fever. Should this primary stage be survived, then, erysipelas, pyæmia, and irritative fever develop themselves, adding their baneful influence to the multitudinous dangers which encompass the sufferer. These facts being remembered, it becomes the duty of the medical man, to attempt some interference by which immediate relief may be afforded to his patient, and amputation and resection become the alternatives which present themselves to his mind. It is important, therefore, to have an accurate knowlodge of the relative value of these two operations, as upon the decision of the Surgeon, human life—the existence of a hero and a martyr—may depend. Relief must come quickly if it come at all. There is no time for delay or investigation when the mutilated victim appeals for succor. The comparative difficulties, dangers and results of the two operations should be fully comprehended and properly appreciated in advance, so that an intelligent response may be made to the demands of science and humanity, without hesitation or delay.

The advantages claimed for Resection may be thus summed up:

1. In the event of a successful issue, the life and limb of the patient are both saved,—the latter anchylosed and deformed it is true, but still not entirely useless.

2. But a small quantity of blood is lost during the operation, and there is no danger of seconda-

ry hemorrhage—an accident which seriously complicates and materially endangers all amputations.

3. There is less of the substance of the limb destroyed, and the shock to the system is not so great as in amputation.

4. In civil practice the results of resection have been comparatively favorable.

The objections urged against the excision of the knee joint, may be thus stated:

1. Even in the event of success, the limb is so completey anchylosed and deformed as to be less useful than an artificial limb of proper construction.

2. Though the danger from secondary hemorrhage is less, erysipelas, purulent infection, excessive and prolonged suppuration, irritative fever, and marasmus, with their attendant evils, are more likely to occur, than after amputations.

3. The convalescence is always tedious, involving a long confinement in the recumbent position, and producing the most serious inconvenience to the patient because of the absolute repose demanded by the necessities of the case.

4. Without the most perfect repose—the absence of all motion, and the most careful after treatment there is danger of destructive inflammation and of great deformity. These constitute the necessary condition in the proper treatment of the case—the *sine qua non* of its management. This fact renders resection of the knee joint in field surgery almost impracticable in view of the means of transporta-

tion, appliances, &c., at the command of medical officers.‡

5. The experience of military Surgeons does not prove this to be so reliable an operation as amputation in the lower third of the thigh. Macleod reports only one case, and that an unsuccessful one in the Crimean campaign. Moreau reports three cases in his experience, all of which proved fatal. The former writer uses in this connexion the following significant language, "Admiring, as I do, the brave attempts which have been made in civil practice to save limbs by excising the knee, I regret that it should not be extended to milita practice; but except in rare cases *I fear that it cannot be accomplished,* from the careful after treatment, and the long period of convalescence necessary to effect a cure."

The Surgical society of Paris has decided positively and unanimously against this operation, in connexion with a case of resection submitted by Maisonneuve. So, likewise, Park declares that this operation indicates "more courage than judgment on the part of the operator;" while Vidal, "in view of the dangers, delays, and bad results" attending it, enters his formal protest against its employment.

---

‡ Erichsen warmly advocates this operation in civil practice, and gives twenty-four cases, of which seventeen were successful and eight died. Ferguson speaks of more than one hundred cases, with a mortality greatly less than that for the thigh. Syme favors the operation and gives numerous instances of its successful employment. In view of these facts, it appears that resection of the knee joint when especially undertaken for chronic diseases of the articulation, or even for wounds when circumstances admit of a proper "after treatment," has been successful in the hands of civil Surgeons.

The conclusions to be drawn from these statements and arguments seem to be plainly these:

1. When the condition of the patient is good, his hygienic surroundings unexceptionable, and the proper means and appliances at hand for the subsequent management of the case, the Surgeon is justified in resorting to it, particularly for disease, and even for accidents.

2. When the condition of the patient is bad, and his hygienic surroundings exceptionable,—as when exposed to the vitiated atmosphere of cities, ill regulated camps, and crowded hospitals—or more particularly, when the circumstances of the case preclude that absolute repose of mind and body so indispensable to its success, the Surgeon is not warranted in attempting the operation.

3. Resections of the knee joint are better suited to civil than to military practice.

4. Resection of the knee joint should not take the place of amputation of the thigh, in the lower third, in field surgery, because of the impossibility of maintaining those conditions which are absolutely necessary to its success.

5. Resection of the knee joint may be resorted to in hospital (military) service, when the tone of the system has not been lowered by exposure, privation, or disease, and an abundance of pure air and nutritious food can be commanded, provided that permanency of location, constant and intelligent attention, and contentment of mind on the part of the patient, can be secured. If there be the slightest doubt or difficulty in regard to either one of

these prerequisites, give the patient the benefit of it, and amputate the limb. Remember that resection of the knee must be performed primarily or not at all.

The drain upon the system is immense, and every possible provision should be made for sustaining and invigorating it at all periods in the history of the case.

Resection of the Hip Joint.—Directions.—Make a semi-lunar incision, beginning at the anterior superior spine of the Ilium, and carrying it behind the articulation to near the tuberosity of the Ischium; cut a large flap with its base downwards through the muscles, and raise it so as to show the capsular ligament of the joint; divide this ligament thoroughly; flex the thigh and carry it inwards; divide the round ligament; carry the knife between the head of the bone, and the acetabulum, and divide the soft parts behind; and then press the head of the bone outwards and remove it with the saw. This being done, bring the flap in position, place the limb on the double inclined plane, or in Smith's anterior splint; and apply cold water dressings. Convalescence is necessarily slow, and is preceded by extensive inflammation, profuse suppuration, and debility.

Remarks.—This operation was first attempted by White, of Manchester, in 1769, and, it is said, with success.

Experience has demonstrated that resection of this joint is much more successful when performed for disease than for injuries; and a rule has been adopted for this special operation among military

Surgeons, which is directly opposed to that established for all others, viz ; for hip joint, resection discard the "primary operation," and rely exclusively upon the "secondary." The most fatal results will, in all probability ensue from haste, while nothing can be lost by delay. *Festinate lente* is the cardinal principle in regard to resections at the coxo-femoral articulation.

This is a dangerous as well as difficult operation and should not be resorted to, save as the alternative of an amputation in the upper third of the thigh, or at the hip joint.

As regards the propriety of this operation, it is well to remark that the sentiments of Surgeons are divided. The following statistical table, taken principally from Armand, will perhaps aid in solving the difficulty, with those who may be called upon to decide this important question.

PRIMARY RESECTIONS AFTER GUN-SHOT WOUNDS.

Surgeons.	No. operated upon.	Cures.	Deaths.
Larrey,	6	0	6
Cooper,	2	0	2
Leteille,	1	0	1
Hutin,	2	0	2
Sedillott,	5	0	5
Sorrell,	1	1	0
Guyon,	1	0	1
Ruchet,	1	0	1
Gibiot,	3	0	3
French crim. service	9	0	9
Macleod,	5	1	4
Stromyer,	1	0	1
	37	2	35

To this frightful record may be added a case recorded by *Seuten*, in 1832, in which death followed the operation. It is plain then that primary operations are to be discarded, and that these are not cases for field surgery.

The statistics of operations for injury show favorable results as compared with amputation of the the thigh near the hip joint, and demonstrate that there are circumstances under which this resection may be properly undertaken. The rule, therefore, is to attempt to save the limb, relying upon a secondary operation, if the effort prove abortive.

The greatest trouble is in the treatment after the operation. It is a matter of prime importance to keep the limb in a state of *repose;* and yet, arrangements must be made to facilitate those movements, which, in the necessary changes of position, are essential to the patient's comfort. Violent extension, then, is both unnecessary and injurious,—unnecessary because it is useless to attempt to restore a perfect limb, and injurious because it prevents those movements which are necessary to comfort and recuperation. To meet these varied indications the double inclined plane, or, better still, the anterior splint of Professor Nathan R. Smith, of Baltimore, should be applied. These appliances will be more particularly described under another head but it will not be inappropriate to say in this connexion, that the latter is one of the great surgical improvements of the present century.

Whatever the nature or extent of the injury, or however great the seeming necessity for this operation, it should never be performed in any Hospital in which pyæmia, hospital gangrene, erysipelas, or cholera prevails as an epidemic, or upon those whose systems are below the standard of health. *

The following principles may be regarded as established in regard to this operation :

1. This operation, though dangerous should be preferred to amputation of the thigh above the junction of the upper and middle third of the femur, or at the hip joint.

2. The secondary rather than the primary operation should be preferred.

3. Nothing is lost by delay, and an attempt to save the limb.

4. Statistics show a mortality after *primary* operations of nearly one hundred per cent, but give somewhat more favorable results for secondary.

5. It is necessary to keep the limb in repose but to provide, at the same time, for the natural and necessary movements of the patient. These two indications are best accomplished by the employment of Smith's anterior splint.

6. Never operate unless all the sanitary conditions are favorable, or when there are difficulties in regard to transportation or subsequent treatment.

7. Sustain the strength of the patient against the immense drain upon his vital resources incident to the profuse suppuration following the operation.

Resection of the Ribs.—It may be necessary to repeat Richerand's operation for resection of a rib, though the Surgeon is seldom called upon to do so.

Directions.—By a straight, a curved, or a conical incision lay bare the diseased portion of the bone; divide the intercostal muscles above and below the rib on a director passed under them; then detach the *pleura* from the bone; and saw through the bone with a chain saw.

There is danger of hemorrhage from the intercostal artery; but the vessel is small and may be readily drawn out and ligated.

Resection and removal of the Inferior and superior Maxillary Bones.—These operations are undertaken for the removal of the principal bones of the face, when attacked by malignant disease; and hence they do not particularly concern the military Surgeon. They are bloody, tedious, and perhaps dangerous, but, both as regards deformity and mortality, their results are far less to be dreaded than is generally supposed, or as might reasonably be expected.

Observations.—In resections art should not only seek to remove the diseased bones but to reproduce the fragments which have been destroyed. That this is possible to a considerable extent, is established alike by clinical observation and the teachings of experimental physiology. A fresh impetus has been given to these investigations by the recent researches of Dr. Leopold Olier, of Paris. His conclusions are of sufficient importance to justify their incorporation into the substance of this volume. The following is the substance of them:

1. The reproduction of bone proceeds from the inner surface of the periosteum.

2. In transplanting portions of the periosteum, bone of various forms and dimensions can be attained according to the shape and position of the transplanted flap.

3. Bones thus developed are not simply shapeless concretions of calcareous matter: they consist of true bone with all the anatomical characteristics of that tissue.

4. The new bone is developed in the subperiosteal blastema, which exists normally upon the inner surface of the periosteum.

5. This blastema consists especially of free nuclei—enclosed in cells floating in a semi-liquid, transparent, or firmly granular material, and mingled more or less with fibrinous elements.

6. The sub-periosteal product which is observed, within the first few days following the transplantation, is generally cartilaginous; but the succeeding development of bone progresses without this intermediate element.

7. An analagous membrane is found after a time upon the surface of the bone from which the periosteum has been removed.

8. When a bone is removed, leaving its periosteum attached to the tissues which ordinarily cover its surface, at the end of a certain time, this portion of the bone is reproduced to a greater or less extent.

9. After resection of the articular extremeties of the two contiguous bones, a new articulation is formed, if the capsule and ligaments are left entire;

while the two long extremeties are remodeled independently of each other.

From these observations, it is conclusively demonstrated, that the preservation of the periosteum is of the highest importance. As bone can be produced in inferior animals, wherever periosteum is transplanted, similar results may be expected in man by retaining portions of the same membrane. After all resections, the excised portions of the bone should be covered with periosteum so as to ensure their speedy union. The apparent difficulties in the way of a practical illustration of these principles should not deter the Surgeon from a persistent effort to adhere to them in as much as they open the way to the accomplishment of such important results in this special branch of Surgery.

These views are new, startling, and in direct opposition of the accepted dogmas of the profession; but they certainly merit attention and consideration, as the land marks of a new field of physiological research, and the heralds of still prouder triumphs for Surgical science.

Let the Surgeon in operating on bony tissue remember, then, to preserve as large a portion of the periosteum as possible, in as much as no possible injury can result from such a procedure, and if the deductions just enunciated be correct, a most important desideratum is supplied thereby. The experiment of leaving the periosteum intact, might, perhaps, be tried to some advantage in connexion with the operation of trephing the skull,—securing a bony covering for the delicate and important parts which are exposed in this operation. For the facts in this regard, collected by my friend Surgeon F. Sorrell, the reader is referred to table "G" of the appendix to this volume.

CHAPTER V.

HEMORRHAGE.

Hemorrhage may be Primary or Secondary, according to the period of its development.

PRIMARY HEMORRHAGE.—A flow of blood may associate itself either with operations or with wounds. When it takes place during the operation, or in a short time subsequent there to, or when it occurs upon the first receipt of an injury or within a few hours after the accident, the hemorrhage is said to be primary.

It is produced, under these circumstances, either by the direct section of the vessel by the amputating knife or the missile causing the wound, and it is instantaneous or delayed according to the extent of shock sustained by the system, or the condition of the artery itself subsequent to the division of its coats.

All Surgeons have observed the fact, that in some instances a division of the large arterial trunks is followed by no immediate loss of blood, and that the flow is not only occasionally delayed, but even entirely suspended. This is observable both in amputations and in wounds, especially of a contused character.

HEMORRHAGE.

This is due to the influence of two causes, which deserve some consideration in this connexion, viz: paralysis of the vessel, and the condition of its internal coat.

1. Paralysis of the vessel.—The influence excited by the nerves upon the circulation, was pointed out in another connexion. It will suffice for the present purposes to state that each vessel is accompanied by nervous filaments, upon the integrity of which the proper performance of the circulatory function depends. When these filaments are so affected by any disturbing cause as to become bad conductors of nervous influence, the flow of blood through the artery is interrupted to a great extent, and even suspended entirely in some instances, notwithstanding the propelling power of the heart which supplies the *vis-a-tergo*. When therefore an artery is severed under these circumstances, or when the vessel is divide by an agency which at the same time paralizes it, there is either no hemorrhage from it, or the blood flows in a very small quantity.

2. The Condition of the Internal Coat of the Artery.—If a cylinder of paper be covered internally with a coat of varnish, and then suddenly and forcibly put upon the stretch, an examination will discover an immense number of points at which this internal coating has been fractured. The arteries are lined with a tunic equally as delicate and friable, and when rudely stretched or torn, as occurs in connexion with lacerated wounds, this internal tunic suffers fracture at a number of points throughout its course. At each point of fracture coagula

tion and effusion takes place, tending to arrest the blood current. This taken in conjunction with the paralysis of the vessel, accounts for the fact that in some instances there is no hemorrhage, even when large arteries are severed, in connexion both with operations and wounds.

Primary hemorrhage frequently relieves itself by inducing *syncope*—a condition in which there is such a stasis of blood in the divided part as admits of the formation of clots and ensures the complete blocking up of its vessels.

The flow of blood may take place from the arteries, the veins, or the capillaries, while the soft parts generally or the bony tissues exclusively may be the seat of the hemorrhage.

The Blood from an artery is of a vermillion color, and flows by jets which are synchronous with the contractions of the left ventricle. It may come either from the proximal or the distal end of the vessel, but generally from the former.

The blood from a vein is of a dark color and flows in a uniform stream. Usually it simply wells out, but when there is pressure as from a ligature, when the position of the part causes the fluid to gravitate towards it, or when the contraction of the muscles constringes the vessel, the blood may be driven out with some force.

The blood from the capillaries is neither so bright as that from the arteries nor so dark as that from the veins, and oozes out rapidly, it may be, but with no force.

Hemorrhage may arrest itself spontaneously, by inducing syncope, or it may cause the speedy death of the patient by depriving the great centres of their "life which is the blood."

If the quantity of bood lost be very great but still not sufficient to produce death, and particularly if it be spread over a considerable interval of time, a state of anæmia will be induced, characterized by pallor of the skin, palpitation of the heart, rushing noises in the head, muscular debility a tendency to syncope, œdema of the lower extremeties, and a general impairment of all the functions.

From this state the patient sometimes rapidly recovers, the vital fluid being speedily reproduced, and the organism readily returning to its normal tone and standard of health.

It not unfrequently happens, however that this state of anæmia becomes the settled habit of the system and continues for a long period, being accompanied by great debility and disturbance of function.

Hemorrhagic fever may also manifest itself after great loss of blood, characterized by a tendency to reaction, with extreme irritability of the heart and arteries. This is nothing more or less than fever associated with anamia, as the symptoms plainly indicate.

Hemorrhage may be delayed until reaction ensues. The current of blood which has suffered a temporary arrest under the shock induced by the injury or operation, may be driven by the more violent contractions of the heart through the vessel notwithstanding the obstructions to its passage, and lost in large quantities cotemporaneously with the development of reaction in the system. This usually occurs within the first thirty-six hours, and may justly be regarded as a primary hemorrhage.

Secondary Hemorrhage.—As that hemorrhage which occurs before the development of inflammation is styled primary, so that which occurs after that process has been established is denominated secondary. It may associate itself either with inflammatory fever, sloughing or ulceration, but as regards the time of its occurrence, is always subsequent to the inflammatory reaction.

It is now generally agreed among Surgeons that any flow of blood which takes place after the thirty sixth-hour succeeding an operation or an injury is to be regarded as a secondary hemorrhage and should be treated immediately as such upon the principles which will be discussed hereafter.

A patient upon whom an operation has been performed, or who has received a wound of any magnitude can never be regarded as beyond the possibility of this accident until the work of cicatrization is complete. From the first cut to the last dressing—at any period in the history of the case—hemorrhage is liable to occur, endangering life and calling for the exhibition of skill and courage on the part of the Surgeon.

The period at which secondary hemorrhage is most likely to occur is still a matter of dispute. Guthrie affirms that it is between the eighth and twentieth day; Dupuytren thinks it is from the tenth to the twentieth day; Henman sets it down as from the fifth to the eleventh; Roux from the sixth to the twentieth; and Macleod from the fifth to the twenty-fifth. It has been known, however, to occur as late as the seventh week, even without the existence of gangrene or ulceration, though after the twenty-fifth day it is fair to pronounce the patient in a great measure over his danger.

This accident may arise from a variety of circumstances, connected with wounds and operations.

In connexion with wounds it proceeds from—

1. The separation of the eschar.
2. Injury by fractured bones.
3. The erosion or tearing of the vessel.
4. Relaxation of the capillaries produced by general feebleness of the patient.
5. Ulceration either incidental or accidental.
6. Gangrene.
7. Development of the collateral circulation and the patulous condition of the distal orifice.

In connexion with operations it may be the product of the following causes:

1. Those which are connected with the condition of the artery.
2. Those which are connected with the ligature itself.
3. Those which are connected with the condition of the blood.
4. Those which are connected with the system at large.

1. *Causes which are connected with the condition of the artery.* The different coats of the artery are subject to diseases of various kinds, and

when so affected, there will ensue rapid sloughing and ulceration of the vessel at the point of ligation, and a consequent escape of blood. Again it frequently occurs that an atheromatous or calcarious deposit has developed itself in the artery, rendering it brittle, and causing it to give way within a day or two succeeding the operation. So, likewise, when the arteries, in common with all other structures of the organism, have yielded to the enervating influences of asthenic diseases, insufficient diet, and such agencies as tend to diminish vital power and to retard nutrition, the ligature readily divides the weakened coats of the vessel, and permits the escape of its contents. The slight wounding of the artery above the ligature, or even of one of its smallest branches, may produce secondary hemorrhage. But the most frequent cause of hemorrhage which manifests itself in this connexion, originates in the patulous condition of the lower orifice of an artery which has been accidentally divided—a condition which results from the division of the nervous filaments distributed to the vessel, and at the same time invites the escape of its blood.

2. *Causes which associate themselves with the ligature* itself. These may depend either upon the nature of the maierial employed, or upon the manner of its application. Thus, as has been already shown, some substances are more irritating to the tissues than others, and by developing too much inflammation in the coats of the artery, cause their disruption, and the development of hemorrhage. Again if the ligature be tied too loosely, or with

the inclusion of a piece of nerve, vein, or muscle, so as to become loose after suppuration has ensued, the blood readily and rapidly escapes. It often happens, that, either from some anomalous development of branches, or the ignorance of the operator, the ligature is tied too near to a collateral branch above, so that the condition of quiescence so essential to the production of a firm coagulum, cannot be obtained, and the plugging up of the artery is not effected. Under these circumstances the blood may escape at any time, causing great trouble and inconvenience, and seriously endangering the life of the patient.

3. *Causes which connect themselves with the condition of the blood.* It has been seen that the formation and organization of a clot—of a firm and adequate coagulum—is essential to the complete and permanent closure of the artery. Physiology teaches that the blood coagulates much better at some periods than at others, and that this difference depends upon certain intrinsic changes which take place in the circulating fluid itself. The blood must contain a certain amount of fibrine and red corpuscles—must be in the possession of its normal and healthful constituents, in order to ensure its ready coagulation, whether within or without the body and it must follow, therefore, that alterations in these elements, both as to quantity and quality, have a material though indirect influence in the development of secondary Hemorrhage. Experience has shown that violent exercise, especially when accompanied by nervous excitement, tends to liquify the blood and to interfere with its coagulability.

Should not this fact furnish a hint to Surgeons, as regards the treatment of wounded arteries, upon the battle field?

4. *Causes which connect themselves with the system at large.* All material changes in the blood either depend upon or induce certain alterations in the system at large; and to that extent this division of the causes which produce Hemorrhage, belongs properly to the last head. There are other states of the system, however, which exercise a more direct influence in the induction of this accident, and which should be considered in this connexion.

It is not only necessary that a firm coagulum should form within the vessel, but the outer coat must be reinforced by a deposit of plastic lymph in order to prevent its rupture. A certain amount of normal and healthful adhesive inflammation is essential to the perfection of this process,—which is impossible in certain diseased states of the system, as when a tendency to erysipelas, phlebitis, suppuration, albuminuria, pyæmia, &c., exists.

Although secondary Hemorrhage may occur at any time in the history of the case, there are *three periods* at which it is particularly likely to be developed, viz: 1 Within a few days after the application of the Ligature, 2 When the Ligature separates; and 3. At an indefinite time after its separation.

1. *Within a few days after the application of the Ligature.*—The bleeding which appears at this period results from the improper tying of the artery; some disease or defect in its coats; from the development of the collateral circulation, and the escape of the blood through the patulous orifice of the distal end of the artery; and from the want of proper adhesive inflammation, &c.

2. *When the Ligature separates and comes away.*—This Hemorrhage may be occasioned by any one of the causes above mentioned; but is mainly due to the improper development of the internal coagulum, and to the absence of the reenforcement which the external coat requires to enable it to sustain the great burden imposed upon it.

3. *After the ligature has separated.*—Hemorrhage may appear at any period between the separation of the ligature and the cicatrization of the wound. This is usually the result of the absorption both of the internal coagulum, and the lymph by which the external coat has been strengthened.

TREATMENT.—The treatment of Hemorrhage consists essentially in preventing or arresting the flow of blood, and is modified by the variety, seat and source of the flow.

Treatment of Primary Hemorrhage.—The means employed for the control of hemorrhage are susceptible of division into two classes, viz: Preventive and Curative measures—the one being employed *in advance* to prevent the flow and the other *after* the appearance of the Hemorrhage, to restrain it.

Preventive measures.—*These* embrace compression, position, and arterial sedatives.

Compression.—The flow of blood can be prevented in most cases by shutting off the supply by means of compression made upon the artery between the locality of the accident and the part. It may be Digital or Instrumental.

Rules for Digital Compression.

1. Find the artery, select the point for compression, and see that the thumb and fingers are applied forcibly upon the vessel.

2. Apply the thumb across the vessel like a seal; or if the fingers be employed form a horizontal plane with their united pulps and range them along the course of the artery.

The thumb is placed upon the opposite side and made to constitute a fixed point upon the limb.

3. Press just hard enough to destroy the pulsation in the vessel, and when the fingers become tired, aid them with those of the other hand.

4. Pressure should be made perpendicularly to the artery.

5. When a jet of blood is required, so as to enable the Surgeon to recognize the vessel, the fingers can be slightly raised, without letting the artery escape, and then reapplied.

The advantages of digital compression are, these viz: the venous current is not arrested; pressure is only made upon one point; and the artery can be always discovered by the sense of touch.

Instrumental compression is accomplished by means of a *key*, the winch, the tourniquet, &c. The key may be employed for compressing nearly all the arteries, especially the subclavian. When used it should be well padded and applied directly across the track of the artery and not too firmly pressed upon it.

The Winch may be used in cases of absolute necessity, but it is objectionable because it compresse the veins, and cannot readily be relaxed or lightened

The Tourniquet of Petit. This consists of three parts, viz: the pad to compress the artery— which should be firm narrow and flat;—a strong band to embrace the limb; and a screw by which this band is tightened, and the artery more firmly compressed. The Pad should be so placed as to compress the artery against the bone; and the screw turned lightly until the first incisions are made, or, what is better still, until hemorrhage from the artery demands some additional assistance for its restraint.

The advantages of the tourniquet are that it can be more readily used by the ignorant—the patient himself being able to manage it properly—; it ensures a more reliable and permanent pressure; it compresses all the branches of the artery as well as the main trunk itself;—it never tires, as do the fingers; it controls hemorrhage as well in anomalous bifurcations and distributions, as under ordinary circumstances, and it presses upon the nerves and thus, to some extent, diminishes the sensibility of the part.

The disadvantages of this instrument are that, it interferes with the venous circulation, and by accumulating blood in the part, causes a great loss of that fluid during an operation; and that it may induce mortification if ignorantly or too persistently employed, by paralyzing the nerves beneath it so as to lower the vital energies of the tissues to which they are distributed, and by cutting off the supply of arterial blood.

The Tourniquet of Signori.—This instrument consists of an arc of steel with a joint in the middle and a screw by which the padded extremities of the instrument are pressed together. One of these pads can be applied directly over the artery, selecting, if possible, some point above the bone and the other on the opposite side of the limb.— By turning the screw the necessary amount of extension is made.

The advantages of this instrument are, that the compression can be rapidly taken from the artery, and that as only two points of the limb are compresssed, the venous circulation is not interrupted.

The objections to it are, that the pad is likely to roll off the artery as the screw is turned, and that in relaxing it, the position of the whole instrument is frequently so much changed as to render a fresh search for the artery necessary.

The ligature *en masse* of Mayor, the compressor of Dupuytren, and other similar instruments are generally abandoned.

Compression of Particular Arteries.—The primitive carotid may be compressed just above the omo-hyoid muscle, against the cervical vertebræ, by means of the fingers, applied perpendicularly.

The facial artery may be readily compressed by the finger on the border of the lower jaw, just in front of the masseter muscle.

The temporal artery may be readily compressed at a point in front of the external ear, two inches from the base of the tragus, by means of perpendicular pressure made with the fingers. Hemorrhage from this artery may be checked by employing a common tailor's thimble and applying a compress over it.

The subclavian artery may be compressed by means of a key or other similar instrument, well padded and applied at a point where the vessel passes over the first rib, just above the clavicle and external to the scalenus muscle. Unless the patient is thoroughly under the influence of chloroform, this procedure cannot be relied on to the exclusion of other measures.

The axillary artery may be compressed under the clavicle, and against the second and third ribs, but a complicated apparatus is necessary, and the difficulties are great. It may, however, be easily pressed against the head of the humerus, by means of four fingers only or with the addition of a cushon. The point of compression is at the union of the anterior and middle third of the axilla.

The brachial artery may easily be compressed by the fingers or tourniquet, against the humerus, at any point along the border of the coraco-brachialis above, and the biceps farther down. There are several important nerves which accompany this artery, and if the pressure is continued too long, the patient suffers great pain. This artery should be compressed in all operations on the upper extremity, below the insertion of the latissimus dorsi muscle, save those of the hand and fingers, and sometimes in those if the pressure made upon the radial and ulna arteries is not sufficient to restrain the hemorrhage.

The radial artery is easily compressed at the

lower third of the fore arm, between the radius and the tendon of the flexor carpi radialis, just where the pulse is felt.

The ulna artery may be reached at the inferior third of the arm, by pressing the flexor carpi-ulnaris against the ulna.

The external iliac may be compressed, in extreme cases, by pressing it against the brim of the pelvis, through the abdominal paricles.

The femoral artery may be compressed in two places, viz: upon *the pubes*, and in the *middle third of the limb.*

This is accomplished upon the pubes, by pushing it forcibly with the thumb or fingers against the pectineal eminence. The pressure should be made *obliquely, upwards* and *backwards*, forming with the horizon an angle of 45°. This compression is safe easy, and much used in all operations upon the *lower extremities.*

In the middle third of the limb it may be readily compressed against the femur, by means of the tourniquet, and even the fingers, taking care to flatten the artery against the bone. This is much used in all operations on the lower extremities save at the hip joint and upper third of the thigh.

The popliteal artery may be compressed opposite the joint, either by means of the tourniquet or the finger

The anterior tibial artery may be compressed by forcing it against the tibia at any point from the middle of the leg to the termination of its course. It is to be found on the side of the extensor proprius-pollicis tendon. Compression of this artery is not of much importance so far as amputations are concerned,—the femoral being compressed in all operations upon the lower extremities.

The posterior tibial may be compressed in the lower third of the leg, at any point parallel with the inner margin of the tendo-Achilis, and also behind the inner ankle, where it is very superficial—not much employed for the reason given above.

Position.—As a means of preventing concurrent hemorrhage, position may be employed to considerable advantage. It is manifest that the normal position is best adapted to the necessities of the animal economy, and that while the course of the principal venous trunks is perpendicularly upwards, the amount of blood carried from the extremities, bears a certain relation to the wants of the various tissues to which the arteries have transported it. A greater amount of the circulatory fluid must therefore remain in the parts concerned so long as this erect position is preserved, than when the vessels are turned perpendicularly downwards, and the *force of gravity* is superadded to the influences which normally operate in returning the blood from the extremities towards the trunk. The same principles apply to arteries, but inversely,—the force of gravity acting as some restraint upon the heart's action, and in a measure controlling the circulation.

These facts may be employed to advantage in the restraint of hemorrhage from wounds; and the elevation of the affected part should be attempted among the earliest measures employed by the Surgeon.

In primary operations also, when the Surgeon has leisure to devote to the work, an attempt should be made to diminish the amount of blood in the condemned member and to relieve the venous trunks and capillaries to some extent, by reversing the position of the part for some moments before the incisions are made. In "secondary amputations," this should invariably be done, as there is always time enough to spare for this procedure, and the hypertrophied condition of the vessels increases its importance.

In all operations upon drunkards this should be an indispensable preliminary on account of the extraordinary stasis of blood in the capillaries of such persons, and the tendency to hemorrhage from that cause. To such a degree does this capillary congestion exist in some cases—and it is present in all—that the simple operation of cupping, or the application of leeches, induces an oozing from these delicate vessels of so persistent a character as to defy all ordinary remedies and to jeopardize the patient's life.

Arterial Sedatives.—These act by diminishing the the amount of blood sent to the part by its propelling organ. To this class belong digitalis, veratrum, viride, tart. emetic, and all those agents which directly or indirectly control the hearts action. They may be employed, with particular advantage, when the necessities of the case demand an amputation and during the existence and manifestation of febrile

ARTERAL SEDATIVES.

phenomena. The subject demands a more thorough investigation at the hands of the profession.

Curative measures.—By this term is meant all those agents which are employed for the arrest of hemorrhage after it has been developed. These vary according to the vessel which is the source of the flow as it comes from arteries, veins capillaries.

Bleeding from the veins, generally ceases spontaneously or is readily controlled by pressure. It is best to avoid applying ligatures to veins when the operation can be avoided, on account of the danger of phlebitis, though it may be necessary when they are diseased or when they have been opened obliquely by wounds or in operations. This variety of hemorrhage results either from mechanical obstruction to the return of the blood to the heart; from the violent struggles of the patient whereby it is prevented from flowing freely through the lungs and larger veins; and from unnatural enlargement of the veins themselves—which induces a retention of blood within them. If the blood flow freely from a large vein during an operation, it may be arrested either by plugging up the vessel with the finger, or applying a sponge saturated with cold water, and taking off the compression above—i. e. relaxing the tourniquet, &c. Should the flow be influenced by the struggles of the patient, a more liberal administration of Chloroform will greatly tend to arrest it, by removing its cause. In the event of a puncture to arrest the hemorrhage by these means course may be had to the various remedies capillaires and arteries, which will be considered at length below.

Bleeding from the capillaires may occur either from the unnatural development of these vessels, from an unusual stasis of blood in them, and from a disproportionate activity between the arteries and veins. These vessels are usually compressed by the retraction of the tissues, but it sometimes happens that the blood continues to flow in streams, greatly to the danger of the patient and to the confusion of the operator.

The means used to arrest the flow of blood from the cappillaires are:

1. Compression by means of the fingers and bringing the flaps together.

2. Cold water applied either by means of a sponge or poured from a vessel.

3. Styptics proper, such as powdered ice, evaporating lotions of water and alcoholic, camphor in powder or between two damp cloths, matico, ergotine, &c.

4. Absorbents—as lint, agaric, spiders web powdered gum arabic, flour, and rosin.

5. Astringents,—including all of vegetable origin, either in solution ˜wder, also alum, sulphate of iron, chloride of i. ·hate of copper, nitrate of silver, vinegar and n juice, creosote and water, sol. sulp' iron, muriated tincture of iron, and com .cture of benzoin.

6. Cauterizati˜ ˜e resorted to when other means have faile 'actual cautery should be used under the˜ ˜tances. Heat the iron to whiteness, and ˜, ˜t a short time to the part˜

taking care not to bring away the eschar, and thus to defeat the objects of its application.

Bleeding from the *arteries* is of most frequent occurrence, as it is the source of greatest danger to the patient. The means at the command of the Surgeon for the arrest of the flow of blood which occurs *during* an operation or immediately subsequent to it, are numerous. The following are most worthy of confidence:

Direct Compression.—This consists in the application of the finger to the orifice of the bleeding vessel until compression can be established above or the operation completed.

Indirect Compression.—This consists in seizing the vessel between the fingers, or grasping it firmly with the hand. This is chiefly practised in flap operations, and especially in amputations at the hip and shoulder joints. The old method of ligating the femoral artery, before commencing the amputation, was always opposed by the best practical Surgeons, and is now considered obsolete,— the method of indirect compression by the hands of an assistant, following the knife and grasping the artery immediately after its division, is now universally recognized as the proper procedure. In operations at the shoulder joint the main reliance of the operator must be upon immediate compression of the artery by his assistant as the last flap is cut, rather than upon pressure above the clavicle. The artery may also be compressed by the application of the ordinary forceps, with hooked extremities and a spring catch, or a forcep with flattened extremities, which closes by its own spring, and is opened by pressing the two blades together.

Ligature of the vessel.—This may be done, by laying the artery bare, applying the ligature, and dividing the artery below it. Again, the two ligatures may be applied, and the vessel divided between them; and finally the artery may be cut, then seized with a tenaculum and drawn out, and the ligature tied above the bleeding orifice. The application of ligatures, will be more freely and thoroughly discussed when the subject of "Hemorrhage after an operation" is considered.

This forcep is applicable to arteries of all sizes, and is the surest of the methods employed for arresting the flow of blood. Tie the main artery first, then find its principal branches, and finally seek out every bleeding orifice. The Ligature was first used by Ambrose Paré in amputations, but the mode of its application has been variously modified by other Surgeons.

Ligatures are *immediate* or *mediate*. To apply the *immediate* Ligature, sponge out the wound well; have the pressure on the artery slightly diminished, so as to permit the blood to flow; seize the artery either with a pair of forceps or the tenaculum; draw it out; and having passed the thread under the instrument, make first a loose knot, then direct the loop over the artery and tie it firmly twice,—an assistant placing his finger on the first knot to prevent its slipping. The instrument may then be withdrawn, and the compression removed to make sure that the artery is completely obliterated. With regard to small arteries, that cannot be readily separated from the soft parts, a portion of the cellular tissue may be included in the ligature with them.

The *mediate ligature* is applied thus. Pass two ends of the ligature through curved needles; push the first into the flesh, at a distance of half a line from the artery, and push it out so as to form a semi-circle; describe a similar semi-circle with the other needle, on the opposite side of the artery; then tighten the ligature and the artery is compressed. This plan has not succeeded well on man.—In the application of ligatures particular care must be observed not to include nerves, and veins.

Torsion. This was pointed out by Galen and renewed by Amusat with great success. It should be employed on arteries of small calibre. *Directions.* Draw the artery out and *isolate* it for half an inch; seize it with a narrow round pointed *forceps transversely* on a level with the wound; and mash it so as to rupture the inner coats, while the proper torsion forceps are applied to the free end of the vessel, and the artery twisted by them upon its axis, from three to eight times. This being done, remove the upper pair of forceps, and sink the twisted end completely into the flesh. *Fricke* simply isolates the artery for half an inch and then twists it completely around for *eight* or *nine* times.

Crushing.—The artery is rubbed and crushed between the blades of the tooth forceps, so as to cause the laceration of the two inner coats. Ledran took the idea from observing that the females of animals beat or crush the umbilical cord with their teeth and that no blood flows after this operation.

Incision, with rupture of the internal coats.—This is done by seizing the artery between two pairs of forceps, one of which is placed transversely, and the other applied lower down in the direction of the vessel. With the lower pair, the two inner coats are ruptured and the fragments pressed upwards in the cavity of the vessel.

Styptics.—The various hœmostatics mentioned under the head of Capillary Hemorrhage, may also be employed to advantage in the bleeding of small arteries.

Cauterization.—The actual cautery only should be used for arteries. Heat the iron to whiteness, and be careful not to apply it for too long a time, lest the eschar be dragged off with it. It is unsafe for large arteries, and secondary hemorrhage from the sloughing of the eschar. Malgaigne however thinks the "iron should only be moderately heated, even below redness, and its application made at very short intervals."

The Seton.—It has been proposed to make two openings in the side of the vessel just above its mouth; and then having folded up the free end of the vessel, to push it into the cavity and to make it protrude on either side between the two slits. This is a tedious and difficult process.

Acupressure.——Dr. Simpson has proposed to arrest hemorrhage by pushing long metallic needles through the integuments, passing them beneath the vessel and bringing them out on the opposite side. In this way the artery is compressed between the needle and the superimposed tissues to such an extent as to arrest its current completely. The vessel soon fills with a firm clot, the blood ceases to flow through it, and the hemorrhage is permanently arrested. When the consolidation is complete, the needles may be withdrawn without detriment to the case. This plan is entirely practicable; and usually ensures a speedy cicatrization, in as much as the needles produce but a slight and transient irritation. Silver needles are particularly non irritating and should be employed, whenever practicable.

Treatment of Secondary Hemorrhage.—Nothing can be more important than a proper comprehension of the principles and means by which this serious accident is to be met and managed. The treatment for this variety of Hemorrhage, as for the primary, may be divided into *Preventive* and *Curative* means.

Preventive means.— As regards the ligature, see that it is of proper material; apply it firmly; and be careful not to include portions of vein, nerve, muscle, &c.

As regards the vessel, seek for a healthy portion; avoid large branches; be particular not to wound the vessel; and, in wounds, never neglect to apply the ligature to both of the divided ends of the artery.

As regards the blood, ascertain its condition, and seek to improve it both by removing the cause of its impairment; and by supplying the deficient elements. Iron, stimulants, good food, and confidence in the Surgeons, are the surest remedies.

As regards the system at large, restrain the force of the circulation by means of Digitalis, and Veratrum Veride, and if necessary treat existing complications. Bring the system to its normal *status*, or as near it as possible.

Curative measures.—By this term is meant those agencies which may be employed for the arrest of Hemorrhage *after* its developement. Should bleeding occur from *a stump after amputation*, the following course should be persued: If the Hemorrhage be developed *only a few days after the operation*, bandage, elevate, and apply cold water; then, if this prove unsuccessful, open the flaps, try cold water, styptics, &c., and bring the parts firmly together; and, if the bleeding still continue, open the wound again, search for the bleeding vessel, and tie it.

If the bleeding occur at a later period, make an effort to arrest it by compression with the horse shoe tourniquet, and if this fail, ligate the artery, if possible, in the wound. When the artery *cannot* be tied in that locality, it must be ligated at the most convenient point above.

Should Hemorrhage present itself *after* a *ligature has been applied*, the Surgeon may adopt the following course: If the artery belong to the trunk and the application of a ligature at a nearer point to the heart be impracticable, an attempt should be made to arrest the flow by means of plugs saturated in certain styptic preparations, as the persulphate or chloride of Iron—the tincture of Benzoin, while an effort is made to restrain the force of the circulation.

If the artery be situated in some one of the extremities, elevation, plugging, and the graduated compress should be employed. When these means fail, and the artery is on the *upper extremity*, the wound should be reopened and the vessel tied both above and below the bleeding point if it be possible; but if not, it should be ligated higher up. Should the hemorrhage still continue, amputate the limb. Secondary hemorrhage, occurring in the lower extremity, is more difficult to control. Tie the artery in the wound, both *above* and *below* the bleeding point; and, when this fails, proceed *at once* to amputate. Experience has demonstrated the impracticability of applying the ligature higher up, as gangrene invariably follows such a procedure.

Bleeding in connexion with a wound. All injuries to arteries threaten to produce secondary hemorrhage, and this danger increases with the size of the vessel. Prevention is therefore the rule of modern Surgery, and it is possible to accomplish this, in a majority of cases, by the proper application of the ligature, in the premises. Make it an invariable rule then, to ligate the artery as soon as the injury has been received, by opening the wound and tying both extremities. If the divided artery be a *small* one, do not disturb it, after it ceases to bleed spontaneously; but, on the other hand, if the *severed trunk be of large calibre*,—as the femoral, the tibial, or the brachial—and no doubt exists as to the nature of the injury, *apply the ligatures*, even *if hemorrhage has ceased of its own accord*. This rule should be particularly observed upon the battle field, as, in the transportation necessary to convey the soldier to the hospital, there is the *greatest possible danger* of a reopening of the vessel, and the consequent destruction of his life. It is impossible to construct ambulances in such a manner as to prevent a great amount of jostling, while conveying the wounded over ordinary roads, especially if the country be hilly, or trains of artillery have passed over them, in advance. This fact should be remembered by the Surgeon to whose care the severed artery first falls upon the field; and, without regard to the mere dictum of recognized authorities, he should follow the guidance of common sense, and employ the only sure means for the prevention of a fatal accident. When the wound is recieved under circumstances which permit the employment of other means of a milder and less heroic character this rule is not so imperative, and position, repose, quiet, compresses, and other prophylactics, may be resorted to in order to prevent the return of hemorrhage.

Bear in mind, that as troublesome, difficult, and dangerous as the operation for the ligation of an artery may be, it is a far *less serious* thing than hemorrhage from a large arterial trunk, such as the femoral, the popliteal or the brachial, and that, where there is a reasonable probability of exposure to such disturbing influences as *tend* to reproduce the hemorrhage, the ligature should be promptly and properly applied, in advance.

If these precautions are neglected, and hemorrhage reappears after having been controlled, it is usually from the *lower portion* of the artery. In such cases the blood does not come in jets, but wells out in a continuous stream, and is of a *dark-*

er color than usual. The course to be pursued is as follows : Bandage the limb from one extremity to the other so as make careful and regular pressure throughout its whole extent; both *above* and *below* the wound, along the track of the main artery, apply a compress saturated with the persulphate of iron, or some other styptic ; elevate the limb ; apply an ice bladder, or a continuous stream of cold water immediately over the wound ; give an opiate, and command absolute quiet. Should there be much force in the heart's pulsations, the circulation may be controlled by the administration of digitalis, veratrum, viride, &c.

If a *second* hemorrhage make its appearance, after the employment of the means referred to above, the artery should be immediately secured, in the wound, if possible, although it be in a condition of profuse suppuration.

If the ligature cannot be applied at this point, tie the artery higher up, according to the rules and principles which will be given in detail, under another head.

When all of these measures have been tried in vain, and the bleeding again makes its appearance, and resists ordinary treatment, *amputate* the limb as a last resort.

Should secondary hemorrhage take place from the veins, in consequence of an opening of their coats by ulceration or suppuration, graduated compression from the extremity upwards, must be employed, and ligation resorted to only in the most extreme contingency.

As this is one of the most fearful and fatal accidents to which the human frame is subject, after the performance of operations on the receipt of injuries, the principles upon which its treatment is based, should be thoroughly understood by the military Surgeon. He must act not only *promptly* but *correctly*, or else lose his patient, and feel himself responsible for the fatal issue. By following the rules established for his guidance in the preceding pages, he will not only have the satisfaction of saving life in many instances, but also of knowing that in any event he has done his whole duty. There surely can be no more comforting reflection than this, to the conscientious Surgeon, amid the cares, responsibilities and discomforts of his arderous life.

Although it cannot be questioned that a ligature applied to the ends of a divided artery is the surest method of arresting the bleeding, it is frequently a matter of the greatest difficulty to find them, or even to determine which artery is bleeding. Again when fracture complicates the wound, great injury may be done by exposing it to the action of atmospheric air. It must, therefore, be borne in mind, that though, ligature of both the proximal and distal end is the rule which should be followed as a general thing, under these circumstances, the main trunk of the vessel should be tied above, according to the plan proposed by Anel, and insisted upon by Dupuytren and others.

CHAPTER VI.

LIGATION OF ARTERIES.

LIGATION OF ARTERIES.—Arteries may be ligated at different points, thus:
1. Above the point of division or disease.
2. Both above and below the point of division or disease.
3. Below the point of division or disease, exclusively.

The circumstances under which the ligation the artery *above the point of division or disease* is demanded, are the following:
1. After amputations for the purpose of arresting the flow of blood.
2. In wounds of small arteries when the hemorrhage cannot be otherwise restrained.
3. In local hypertrophies for the purpose of arresting the nutritive process by withholding the pabulum supplied by the blood.
4. In connexion with malignant tumours and for the purpose of restraining their development.
5. In aneurismal tumours, according to the

teaching of Hunter, taking care to expose the artery at some distance from the seat of disease.

6. In wounds of large arteries when it is impossible to ligate both the proximal and distal end.

7. In hemorrhage from an artery in simple fracture, performing Anil's operation according to the views of Dupuytren.

8. In secondary hemorrhage of an uncontrollable character from stumps, &c.

9. In violent inflammations of articular surfaces, when neither resections nor amputations are admissible.

The circumstances which demands the ligation of the artery, both *above* and *below* the point of division or disease are the following:

1. In secondary resections where the collateral circulation has been developed, and the hemorrhage is excessive,—the operation being tedious and prolonged.

2. In traumatic aneurisms, particularly those of the artero-venous variety, the ligature should be thus applied. The older Surgeons treated all aneurisms in this method, but it is now limited to those of traumatic origin.

3. In wounds generally when an artery of large size is divided, as a means of preventing secondary hemorrhage.

4. In secondary hemorrhage when from the dark hue of the blood, and the continuity of the stream, it is plain that the blood issues from the distal end of the artery. The application of the ligatures both to the proximal and distal ends of the artery, under these circumstances will be readily appreci-

ated when it is remembered that the latter does not close as does the former, and that, as a natural consequence, so soon as the collateral circulation is developed, the blood comes welling up from the patulous orifice in obedience to the physical law which constrains a fluid to seek its own level under all circumstances. The causes which prevent the closure of of the distal end, depend for their operation upon the division of the nerves distributed to that portion of the vessel, and the retention in it, immediately subsequent to the operation, of too small a quantity of blood to ensure the formation of a clot sufficiently large and firm to block up the vessel. This method of guarding against secondary hemorrhage, and of restraining it when developed, has become one of the axioms of modern surgery, and should be incorporated into the professional creed of every medical man as a cardinal principle. The neglect of this most simple but significant precept may induce fatal results, for which the Surgeon alone should be responsible, whatever of mortification to him or disgrace to the profession, is incurred thereby. As before remarked, it is not always possible to ascertain from what artery the blood comes, or to find the severed ends of the bleeding vessel; but the operation should not be abandoned for any other, until a diligent search has been instituted and an intelligent effort made to fulfill the indication of the case, in the manner referred to above.

The blood from the distal portion of the divided artery may be recognized in the lower extremities by the darkness of its hue, but in the upper extremities both ends bleed scarlet blood because of the free anastamoses of the vessels.

It is not so important to secure both ends of the smaller arteries, as they can be more readily obliterated, if necessary, or controlled in any event.

The circumstances under which the artery is ligated *below the point of division or disease* exclusively are as follows :

1. In aneurisms of large vessels when the Hunterian operation has failed or is impossible.

2. In aneurisms when the coats of the artery are diseased in consequence of calcarious or artheromatous deposits.

3. In wounds when the hemorrhage is of a dark character and comes in a continuous stream, and the upper portion of the divided artery has retracted beyond the reach of the Surgeon, or is in such close proximity with important organs as precludes its seizure without serious injury to them.

Brasdor proposed to cure aneurismal tumours by ligating the artery only on the distal side, expecting thereby to retard and diminish the current passing through the tumours to such an extent as to ensure the consolidation of its contents. Experience has shown that the Hunterian method is far preferable, and that the procedure of Brasdor is a senseless substitution save in those cases where from the peculiar surroundings of the vessel the former cannot be performed. Wardrop supposed that by tying the artery on the distal side, but beyond a point of bifercation, that the conditions most essential to solidification of the aneurism would be secured. The incorrectness of his views in this regard is demonstrated by the universal abandonment of his operation.

Structure of Arteries.—It is important to understand the anatomical structure of arteries before entering upon the consideration of the general rules for the application of ligatures.

Arteries are tubular vessels of cylindrical form, dense in structure, and composed of *three coats*, the *internal, the middle* and *the external.*

The internal coat is elastic, and composed of two layers, the *innermost* one being only a layer of epithelial cells, resting upon an elastic, but extremely thin, brittle, transparent and colorless membrane.

The middle coat is composed both of muscular and elastic fibres, being highly elastic, and of a *reddish yellow* color. These muscular and elastic fibres are arranged in layers, *encircling* the vessel, and therefore, admitting of an easy division of this coat, under the presure of a ligature applied in the same direction.

The external or elastic coat consists of condensed areolar and elastic tissue. In large arteries the elastic tissue forms a distinct layer, the fibres of which run longitudinally, while another layer of condensed areolar invests the whole,—its fibres being disposed more or less obliquely or diagonally around the vessel.

The arteries are included in a thin areolar investment known as the *sheath*, and are supplied with blood vessels and nerves like other organs of the body; while they are accompanied by *satellite veins*, called *venæ comites*. The nutrient vessels arise from the main artery, from some of its branches, or from a neighboring vessel, and are distributed to the external and middle coats, and possibly to the internal, also. Minute veins serve to return the blood from the vessel into the *venæ comites*. The veins are derived principally from the sympathetic, and par*t*ly from the cerebro spinal system,—forming intricate plexuses upon the surface of the larger trunks, while the smaller branches are accompanied by single filaments.

These vessels are named arteries from two Greek words signifying "to contain air," from the ancient popular but most mistaken ideas respecting their functions.

The action of Ligatures.—When a ligature is tightly applied to an artery of considerable size, certain pathological phenomena are developed, worthy the faithful study of the Surgeon. These effects occur in the following order: An immediate division of the internal and middle coats—the external remaining in tact;—these coats retract and contract forming a cul-de-sac, at the bottom of which, there is first deposited a small nodule of lymph of a yellowish or buff color; this coagulum assumes a conical shape, its base being downwards and is c posed of exudation matter and fibrin closely   herent to the lower end of the artery, while it pex is pointed upwards, floats loose in the ves   l and is composed of fibrin, of a dark purple   maroon color; about the tenth day the plastic ly ph, thrown out in consequence of an inflammat n from the divided coats, binds them firmly to the inclosed plug, the darker portions of which begin to disappear; the vessel contracts still more, and the absorption of coloring matter continues, until the base of the plug becomes incorporated with the contiguous arterial coats and is finally transformed into fibro—cellular tissue. In the *external* coat a certain amount of inflammation is induced by the pressure of the ligature, and plastic lymph is exuded between the vessel and its sheaths which finally organizes and materially strengthens the artery immediately contiguous to the noose as well as over it. The ligature finally ulcerates through the vessel, and its place is still farther supplied by deposits of plastic matter upon the external coat of the vessel.

It will be seen therefore that the simple retraction and contraction of the severed coats, together with the formation of a coagulum, are not sufficient to secure the occlusion of the vessel, but that the inflammatory process, accompanied by effusion of plastic lymph, must develop itself in order to effect the desired result. The delicacy of the arterial coat ensures the induction of this inflammation when the ligature is applied under ordinary circumstances,—a provision of immense importance to the Surgeon, and seemingly designed with especial reference to the success of the art.‡

The instruments and appliances required for this operation are few and of simple construction.—

---

‡ There may be too much or too little inflammation,—the one sometimes resulting in the breaking down of the coagulum by suppuration, the other causing the exudation of so little fibrine as to preclude the ormation of a sufficiently firm clot to ensure obliteration.

Thus, the Surgeon should always be provided with a bistoury, a grooved director, forceps, aneurismal needles, blunt hook, tenaculæ, ligatures, suture needles, adhesive straps, chloroform, styptics, cold water and brandy.

The objects to be held in view in the performance of this operation are three in number, viz :

1. To expose the sheath of the vessel.
2. To isolate the artery.
3. To place the ligature around the artery.

Uncovering the Artery.—The general rules in this regard, may be summed up thus:

1. Make sure of the position of the artery by understanding the anatomy of the part, causing the muscles to contract, feeling the pulsations, and "make assurance doubly sure," by marking out, upon the limb, the exact course of the vessel.

2. Make the skin tense without altering its relation to the artery; and if the vessel be *superficial*, cut *directly* through the skin and parallel with it; but if it be *deep* divide the skin *obliquely*.

3, If the artery lies directly under the *superficial fascia*, or aponeurosis, these should be opened at the side of the vessel to avoid puncturing it; but if the artery be deep they should be opened directly above it. Should the artery not be seen after these incisions, make the muscles contract, and separate them at their *insterstices*, by means of the director or the handle of the knife. When the *deep* aponeurosis is exposed it should be divided according to the directions given for the *superficial*.

4. The artery may be recognized by its pulsa-

tions, by its being thicker than the veins, and by its dull white color.

5. However superficial the artery, *two* incisions are always necessary to uncover it—the *skin* and the *aponeurosis* must always be divided, and by separate cuts. The Surgeon should never cut blindly, but always with a definite object in view, and with a full knowledge of what he is doing.— He should have certain anatomical land marks to guide him to the attainment of his object, and should content himself with quietly finding each in its turn until the goal is reached, without seeking to attain it at a bound, or by an extemporized "short cut."

The Isolation of the Artery.—The rules for the guidance of the Surgeon in separating the artery from its surroundings are as follows:

1. Hold aside the lips of the wound, and remove all pressure upon the artery so as to distinguish its pulsations, and when the sheath is fairly exposed, and opened, pass in the grooved director, and enlarge the opening either by cutting or tearing the membrane. Then, separate the artery from its accompanying veins and nerves, and pass the grooved director beneath, and thus isolate the vessel.

2. If the artery be small, or yellow—indicating disease—its sheath should not be opened. If it be large, open the sheath, separate it from "venæ comites" and "satellite nerve;" and *fix it steadily with the finger and thumb* so as to pass the director under it.

3. If important parts be taken up with the arte-

ry by the director, use another to ensure its more complete isolation.

4. Be sure that you have tied the artery. Some have tied important nerves instead of the vessel,— with the most destructive consequences to the patients, and to their own reputations. Take pains therefore, to feel the pulsations of the vessels before ligating, and to ascertain that the current of blood has been arrested by the operation, by examining the artery both *above* and *below* the ligature.— With regard to the veins, their colour will prevent mistakes

Applying the Ligature.—The rules for applying the ligature are as follows,

1. The ligature must compress the artery perpendicularly; if placed obliquely, it will slip and not sufficiently compress the vessel.

2. The ligature should neither be too small, nor too loose, but should vary according to the vessel ligated, having a certain relation to the size of the artery. Thus, the femoral should be tied with a larger thread than the facial, and so on for the rest.

3. Do not tie an artery immediately below a branch

4. Disturb the ligature, after it has been adjusted, as little as possible.

When an artery is *diseased* or *brittle*, the ligature should be large, and tied loosely.

For the other facts in regard to the application of ligatures, the reader is referred to the previous chapter on hemorrhage.

Treatment.—After the operation has been performed the limb should be placed in such a position as will permit the blood to flow readily from it, while the muscles are relaxed and the lips of the wound are neither patulous nor puckered. The wound should be closed with adhesive straps; the ligatures brought out of its upper portion; a light roller bandage applied; and the cold water treatment instituted. Provision should also be made for preserving the vital warmth of the limb, by wrapping it in flannel, laying it in a bed of soft wool or cotton, using friction and employing artificial heat if necessary. The ligatures should not be touched for *eight or ten* days, if the artery be small, and for *two weeks or more* if it be of large calibre. If symptoms of plethora appear from the mass of blood being confined within more circumscribed limits, blood letting and the usual antiphlogistic treatment should be resorted to without delay. Should gangrene result from the ligation of the artery, amputation offers strong hope for the patient, and it should be employed without hesitation or delay. This accident is particularly likely to appear in connexion with extensive gun shot wounds, or when owing to the ignorance or carelessness of the operator, the large conducting vein from the limb is injured, or an aneurismal communication is formed between the artery and its accompanying vein. Should hemorrhage occur as the ligature separates, compression may be tried, and if this fail, another operation resorted to as the surest means of arresting the flow.

LIGATION OF PARTICULAR ARTERIES.—Under this head will be considered the rules for the ligation of the arteries of the trunk, and of the superior and inferior extremities.

ARTERIES OF THE TRUNK—*The Arteria Innominata.* This artery is the first large trunk given off from the arch of the aorta, and ascends obliquely on the right side, to a point opposite the articulation of the clavicle with the sternum, where it termidates by dividing into the subclavian and common carotid. It is about one inch and a half in length, in the adult, and is in front of the trachea.

PLAN OF RELATIONS.

In front.—The sternum, sterno-hyoid, and sterno-thyroid muscles, remains of the thymus glaud, left innomenata and inferior thyroid veins.

Right side.—Right Vena innomenata, right pneumogastric nerve and pleura.

Left side.—Remains of the thymus gland, and left carotid.

Behind.—The trachea.

Operation.—Directions.—Place the patient in a recumbent position, with the neck slightly flexed and supported with a pillow,—the face being turned in an opposite direction, so as to relax the sterno-cleido-mastoid muscle. Standing upon the *right* side, make a transverse incision, three inches long, commencing at the *median line* of the neck and extending outwards parallel with the clavicle but half an inch above its upper border;—then make another incision of the same length along the inner border of the sterno-cleido-mastoid, terminating at the commencement of the first; open the platysmar muscle and superficial fascia carefully, so as to expose the *sternal* portion of the sterno-cleido-mastoid; divide this muscle upon the *grooved director;* separate the clavicular origin of the muscle upon the inner side of two thirds of its length and reverse it *upwards* and *outwards*; next divide the sterno-hyoid and thyroid muscles, cautiously upon the *grooved director;* open the cellular tissue lying above the vessel with the *finger* or *director* avoiding the right *internal jugular* vein, which is only a quarter of an inch on its OUTER side, and the inferior thyroid veins which cover it in *front* and are to be drawn off on one side;—find the common carotid first, and trace it down with the finger until the innominata is discovered; separate the vessel carefully from the vena innominata on its outer side, and press it off from the laryngeal; and then pass the ligature under it by means of a curved aneurismal needle *from without inwards*. The longitudinal incision may be made first, and perhaps it is more convenient to do so, as the skin becomes relaxed after the transverse one is made. The parts should be brought together and cold water dressings applied.

The propriety of attempting this operation under any circumstances is very doubtful though the facts of the accidental obliteration of this artery demonstrates the possibility of success.

Ligation of the Common Carotid Artery

PLAN OF RELATIONS.

In Front.—Integument, fascia, platysma, sterno-mastoid, sterno-thyroid, omo-hyoid, descenden noni nerve, sterno-mastoid artery, superior and mid: thyroid veins, and anterior ugular.

Externally.—Internal jugular vein and pneumo-gastric nerve.

Internally.—Trachea, thyroid gland, recurrent laryngeal nerve, inferior thyroid artery, larynx and pharynx.

Behind.—Longus-colli, sympathectic nerve, rectus anticus muscle, inferior thyroid artery and recurrent laryngeal nerve.

The common carotid arteries extend *from* a point opposite the articulation of the clavicle and sternum *to* a point on a level with the superior margin of the thyroid cartilage, where, they divide into the external and internal carotids. Both arteries incline backwards as they ascend, while the right is shorter than the left, and somewhat more superior, in consequence of its coming off from the innominata. Each artery is invested in a sheath which contains also, the par vagum nerve and the internal jugular vein—the artery being on the *inner* side, next to the trachea—the vein on the *outer* side, and the nerve *between* the two but a little *posterior* to them.

The place of election is immediately below the bifurcation of the vessel, opposite the thyroid cartilage, and *above* the omo-hyoid muscle.

The place of necessity is anywhere *below* the omo-hyoid and in the inferior triangle of the neck.

Directions for the operation at the place of election. Place the patient in a recumbent position, with his face turned to the opposite side, well supported by an assistant, and his chin carried back so as to extend the integuments in front of the neck. Make an incision on the anterior edge of the sterno-cleido-mastoid, beginning an inch below the angle of the jaw and extending half-way down the neck: raise and divide, on the grooved director, the platysma muscle and superficial fascia, avoiding the anterior jugular vein and the superficial nerves; divide, in the same manner, the deep layer of fascia, connecting the edge of the sterno-cleido-mastoid to the sterno-thyroid and hyoid muscles; lay down the scalpel, lower the chin to its usual position so as to relax the muscles, and hold the margins of the wound asunder with blunt hooks or the fingers of an assistant; then, with the point of the director, the handle of the knife, or the finger, break up the cellular tissue so as to expose the sheath of the vessel, on which is the descendens noni nerve; raise the sheath carefully with the forceps, and open its *inner side*, and enlarge the orifice on a director so as to expose the vessel; hold the internal jugular vein slightly *downwards* and *outwards*, isolate the artery, and pass the ligature under it, by means of an aneurismal needle, from *without inwards*. If the internal jugular vein should by any accident be severed in the operation pass two pieces through its edges and across the orifice, and immediately apply a ligature both above and below the bleeding point. Bring the wound together and dress according to the usual rules.

Directions for the operation at the point of necessity, *below* the omo-hyoid muscle.—Make an incision three inches in length along the inner margin of the sterno-cleido-mastoid terminating at the top of the sternum; an inch from this point, make another incision parallel with the clavicle—ending just beyond the sterno clavicular articulation: divide the sternal portion of the muscle and turn it backwards; and then proceed to isolate the artery and to apply the ligature as directed under the last head.

According to Norris the carotid artery has been

ligated 149 times, and with a fatal result in 32 cases. The most common cause of death after this operation is a cerebral disturbance, which fact can be readily understood when the pathological susceptibilities of the brain are taken into the account together with the important functions of the carotid as the great blood carrier to that delicate organ. Erichsen gives the following as his conclusions in regard to this operation:

1. Ligation of one carotid is followed in about one fifth of the cases by cerebral disturbance, more than one half of which are fatal.

2. Ligation of both carotids at *the same time* invariably results in death.

3. When both carotids are ligated, with *an interval of some days*, there is not more danger than when one is tied.

4. Pathological investigation has shown that even if both the vessels be gradually obliterated the patient may live.

Jobert and Miller have also called special atten- to the fact that the lungs are secondarily affected after the ligation of the carotids.

When the carotids are ligated the head is supplied with blood by means of the vertebral arteries, and a communication which exists between the arteria—princeps cervicis a branch of the occipital, and the profounda cervicis, a branch of the subclavian.

Ligation of the External Carotid artery.—The common carotid of either side divides into the external and internal carotids nearly on a line with the *upper border* of the thyroid cartilage.

The external at its origin is slightly in *front* and

to the *inner* side of the internal carotid, and may be found without much difficulty, by tracing up the course of the common carotid with the finger. Both the external and internal are sufficiently superficial to be readily reached, by the Surgeon; but the latter is not a proper subject for operation for many obvious reasons.

The external carotid has numerous and important branches conveying blood to the thyroid gland, tongue, pharynx, face, posterior aspect of the head, anterior and middle portion of the scalp, carotid gland, &c.

PLAN OF RELATIONS.

In front.—Integument, platysma, superficial fascia, deep fascia, hypoglossal nerve, lingual and facial veins, digastric and styto-hyoid muscles, facial nerve, parotid gland, tempral and maxillary veins.

Internally.—Hyoid, pharynx, parotid gland, ramus of the jaw.

Behind.—Superior laryngeal nerve, styloglossus muscle, styto-pharyngeus and glosso-pharengeal nerves, and paroted gland.

It is only in the cervical portion that the artery is tied, just below the digastric muscle. Above that locality the operation becomes much more difficult and dangerous because of the important parts with which it is in immediate relation.

Directions.—Make an incision, commencing half an inch below the angle of the jaw and extending as low as the middle of the thyroid cartilage and running parallel with, and half an inch from the edge of the sterno-cleido-mastoid; divide the platysma and cervical fascia on a grooved director; separate the sheaths of the submaxillary upwards and forwards; lay bare the digastric and stylo-hyoid muscles at the bottom of the wound, by means of the point of the director or the forceps and draw them forward with a blunt hook: hold the sides of the incision wide apart, carry the nerve and vein backward with the end of the finger, and cautiously open the sheath of the vessel; and then, with the artery isolated apply the ligature by means of an aneurismal needle. Dress in the usual manner.

The external carotid has been tied successfully for wounds, for aneurismal enlargements of its branches, in resections of the jaws, and for tumours of the antrim, and for removal of the parotid gland. Except for wounds which divide it, there is much doubt as to the propriety of the operation, on account of the secondary hemorrhage which almost necessarily follows the ligation of a large artery so near its point of ramification; and with such extensive anastomosies.

Ligation of the Superior Thyroid artery.—It is only necessary to remark in regard to this artery that from its position on the neck, it is divided generally in abortive attempts at suicide, and hence, the only operation necessary is simply one for securing cut extremities in the existing wound. Should it be impossible to do this in consequence of the effusion of blood in the surrounding cellular tissue, and the heaving motion incident to respiration, ligation of the common carotid becomes necessary.

Ligation of the Lingual Artery.—This is a branch of the common carotid and is given off a little above the superior thyroid, from whence it runs to the tongue.

It should be ligated just opposite a small osseous projection upon the upper border of the great cornu of the os-hyoides, one or two lines from the lesser cornu.

Directions.—Find the great cornu and make an incision about an inch and a half in length through the skin and platysma, two lines above and parallel with it; push up the sub-maxillary gland and find the tendon of the digastric muscle, and the hypoglossal nerve; free this nerve and divide the muscle; open the sheath of the artery; isolate and ligate it.

This is a difficult and dangerous operation; but it may be undertaken in wounds and in operations on the tongue. Amussat and Mirault tied this artery for the purpose of arresting cancer of the tongue—a most dangerous and useless undertaking.

Ligation of the Facial Artery.—This is given off just above the lingual and sometimes by a common trunk with it. It runs over the jaw at the anterior border of the masseter muscle, where its pulsations can be distinguished. It is covered by the integument, platysma, and cellular tissue, the facial vein being on its temporal side and some branches of the facial nerve running across it.

Directions.—Make an incision one inch and a quarter across the jaw-bone at the anterior edge of the masseter muscle; open the cellular tissue on the grooved director, avoiding the branch of the facial nerve; open the sheath; isolate and ligate the artery.

Ligation of the Subclavian Artery.—The subclavian of the right side arises from the arteria innominata, opposite the articulation of the clavicle with the sternum, and extends to a point just below the margin of the first rib. On the left side the subclavian rises directly from the arch of the aorta, and is, consequently, longer than the other, and more deeply seated. It follows therefore that the two vessels must, in the first portion of their course, differ in their length, their direction, and their relations with neighboring parts. As a means of facilitating the study of this vessel, especially in a surgical point of view, the subclavian has been divided into three parts. The *first* portion is included between the origin of the artery and the inner border of the scalenus anticus muscle; the *second* is immediately behind the scalenus anticus extending from the inner to the outer border of that muscle; and the *third* extends from the outer margin of the scalenus to the lower border of the first rib.

In its *first* portion, the course of the right artery is obliquely upwards and outwards; in its *second*, it is transversely outwards; and in its *third*, obliquely downwards and outwards, so that it forms, between its terminal points, an arch whose centre is nearly behind the scalenus anticus muscle.

The left artery passes almost perpendicularly upwards to the scalenus muscle and then curves outwards and downwards to the lower border of the first rib. These three portions will be considered separately.

RELATIONS OF THE FIRST PORTION OF THE RIGHT SUBCLAVIAN.

In front.—Integument, superficial and deep fascia, platysma, sterno-mastoid, sterno-hyoid, and sterno-thyroid muscles, internal jugular and vertebral veins, pneumogastric, phrenic and cardiac nerves.
Behind.—Recurrent laryngeal, and sympathetic nerves, longus-colli, and transverse process of the seventh cervical vertebra.
Beneath.—The pleura.

The relations of the first portion are not interesting in a surgical point of view since the artery cannot be ligatured on account of its great depth and close connexion with the pleura.

On the right side the operation has been performed with success; but it should never be undertaken when it is possible to ligate the artery either in its second or third portion.

Directions.—Place the patient upon the table in a horizontal position; make an incision along the inner border of the clavicle; make a second along the inner border of the sterno-cleido-mastoid, meeting the first at right angles; divide the sternal attachment of the muscle and turn it outwards; cut through a few small veins, and divide the sternohyoid and thy-roid upon a grooved director, in the same manner, occasionally the anterior jugular is cut in this step of the operation; cut through the deep fascia with the finger nail and expose the internal jugular vein, which crosses the artery; press this aside and secure the artery, by passing the needle from *below upwards*, so as to avoid injury to the pleura.

Take care to avoid the recurrent laryngeal, the phrenic and sympathetic nerves, and to apply the ligature near the vertebral artery, so as to secure as much room as possible for the formation of a clot.

Aneurisms of the axillary or subclavian artery encroaching upon the scalenus muscle, or wounds of the second portion of the artery, *may possibly* justify this operation as a last resort. This is howeever a tedious, difficult, and dangerous procedure, and should not be attempted without due consideration and for the most cogent reasons.

PLAN OF THE RELATIONS OF THE SECOND PORTION OF THE ARTERY.

In front.—Platysma, sterno-mastoid and scalenus anticus muscles; phrenic nerve and cervicle fascia.
Above.—Brachial plexus and omo-hyoid.
Below.—First rib.
Behind.—Scalenus medius muscle.

Directions.—Place the patient upon a table, secure his head, and see that his shoulders are drawn downwards and slightly forwards ; make an incision immediately above the clavicle and parallel with its posterior border, commencing one inch above the sternal end of the bone, and dividing the external fasciculus of the sterno-mastoid; find the tubercle on the rib with the finger; pass a director behind the scalenus and divide it thoroughly ; then the artery being exposed and recognized by its general course and pulsations, pass the needle from *without inwards*, and apply the ligature. Be careful not to injure the phrenic nerve, the internal jugular vein, and the internal mamary artery whic descends on the inner side of the scalenus anticus muscle.

This portion of the artery, though more favorable for the application of a ligature than the *first*, is far from being the most desirable position for the operation, because of the intimate relation of the phrenic nerve, the internal jugular vein, and the internal mammary artery with the scalenus muscle which must necessarily be divided. There is also another objection which is based upon the close proximity of the artery to the pleura,—a structure of peculiar delicacy of organization. Sometimes the artery passes in front of the scalenus muscle, and occasionally through its fibres.

PLAN OF THE RELATIONS OF THE THIRD PORTION OF THE ARTERY.

In front. Integument, fascia, platysma, external jugular, supra scapula, and transverse cervical veins, cervical plexus subclavius muscle, supra scapular vessels, and clavicle.
Above.—Brachial plexus, and omo-hyoid.
Below.—First rib.
Behind.—Scalenus medius muscle.

This is the most eligible position for the performance of the operation.

Directions.—Place the patient upon a table with his shoulders depressed and his head well secured; draw down the integuments as much as possible, upon the clavicle; make an incision through the skin, thus drawn down, to the bone from the anterior border of the trapezius to the posterior border of the sterno-mastoid; make a short vertical incision meeting the centre of the preceding one at a right angle; divide the platysma and superficial fascia upon a grooved director; hold aside the internal jugular vein, which is on the *inner* side, as well as the scapular and transverse cervical; avoid the supra-scapular artery, and find the omo-hyoid muscle, and hold it out of the way; divide the fascia with the finger nail or scalpel and find the outer margin of the scalenus anticus; and then pass the finger down this margin until it strikes the first rib, where the pulsations of the artery may be felt, as it passes over its surface. This being done, pass the aneurismal needle around the vessel from *before backwards*, taking care not to include a branch of the brachial plexus in the ligature. Remember that the subclavian vein passes almost transversely forwards from the outer margin of the first rib to the sterno-clavicular articulation, in front of the artery, being separated from it by the scalenus anticus muscle and the phrenic nerve.

Remarks.—That portion of the artery which is included between the outer margin of the scalenus muscle and the lower border of the first rib, is always selected as the proper site for deligation, when it is possible to do so. The artery in its third part is comparatively superficial, whilst it is most remote from the origin of the large braches, and not so completely environed by important vessels and nerves.

This operation may be required on account of aneurisms or wounds of the axillary artery; and though less difficult than those undertaken at the *first* and *second* portions of the vessel, it is of sufficient gravity to preclude its employment save in cases of paramount necessity.

In ordinary cases the artery is not at a great depth, but when the clavicle is elevated from the presence of a large aneurismal tumour, it is then very remote from the surface, and the difficulties of the operation are increased.

The circulation of the limb is supported after ligature of the subclavian, principally by means of the superior scapular artery.

In persons with short necks the first rib is lower in relation to the clavicle, and the artery is deeper while the very opposite of this is true in persons with long necks.

The artery is found invariably on the outside of the *projecting tubercle* or the first rib, which gives attachment to the scalenus anticus muscle.

Ligation of the Common Iliac Arteries.—The abdominal aorta bifurcates opposite the body of the fourth lumbar vertebra on the left side of the spinal column and forms the common iliac arteries.— These are about two inches in length, and diverge on either side, running downwards and outwards upon the margin of the pelvis, and dividing opposite the articulation of the sacrum with the last lumbar vertebra, into the external and internal iliac arteries. The external iliacs are distributed to the inferior extremities while the internal iliacs supply the viscera and parietes of the pelvis.

The right common iliac is longer and more oblique than the left. *In front* it is covered by the peritoneum, the intestines, and the branches of the sympathetic nerve, while it is crossed at its dvision by the ureter. *Behind* it is separated from the last lumbar vertebra by the common iliac veins. On the outer side it is in the relation with the vena cava the right common iliac vein, and the psoas magnus muscle. The commencement of this vessel corresponds with the left side of the umbilicus on a level with a line drawn from the highest point of one iliac crest to the opposite one, and its course to a line extending from from this point downward towards the middle of Pouparts ligament.

Directions.—Make an incision from four to five inches in length, from about two inches above and to the left of the umbilicus, outwards in a curved direction, towards the lumbar region, terminating a little below the the anterior superior spine of the ilium; divide carefully each abdominal muscle, and the *transversalis* fascia at the lower part of the wound; separate the peritoneum, together with the ureter, from the transversalis and iliac fascia, and push it well aside; turn the patient on the sound side, and find, with the finger, the sacro-iliac articulation, over which the pulsations of the artery may be felt; expose the artery, together with its accompanying vein, which is in the sheath and on the inner side; isolate the artery and pass the ligature under it from *within* outwards.

If the *iliac region* be selected for the operation, make a curved incision about five inches in length, commencing on the left of the umbilicus, and carried, first outwards towards the anterior superior spine of the ilium, and from them along the upper border of Poupart's ligament to its middle and then follow the directions given above.

Remarks.—This operation has been performed with success though it is, of course, both difficult of execution and dangerous in its consequences. The indications for its performance are, aneurisms, wounds, involving the external and internal iliac arteries, or secondary hemorrhage after amputation of the superior third of the thigh.

It is of the first importance to avoid wounding the peritoneum, lest inflammation be developed in that delicate and susceptible membrane, and thus add another source of danger to the patient's life. It should be carefully held aside, by the finger or a copper spatula in the hands of an assistant, and most tenderly handled.

According to Quain the length of the vessel varies greatly,—ranging in five sevenths of the cases between one and and a half and three inches.— When the artery is found to be very short, it is better to tie both the external and internal iliacs below.

The points of importance are the relations of the vessel to the lumbar vertebra, to the crest of the ilium, to the umbilicus, to the vena cava and common iliac veins of the right side, and to the inner side. In making the incision, care must also be taken not to carry it too low down or too far forwards, as in doing so there is danger of wounding the epigastric, and circumflex-ilii arteries.

Of seventeen cases referred to by Erichsen, nine recovered,and eight died. In two of the fatal cases the peritoneum was opened, and in four of the others, death seemed more the result of the original affection than of the operation. When the depth of the artery is considered, together with its great size, the force of the blood current through it, the intimate relations sustained by it to important structures, and its proximity to the heart, the dangers and difficulties of the operation must be sufficiently patent to inspire the Surgeon with caution and apprehension in regard to it, notwithstanding the statistical information furnished by Ericksen and others in this connexion.

Ligation of the Internal Iliac Artery.—The internal iliac artery is a short and thick vessel which commences at the bifurcation of the common iliac, and, passing to the margin of the greater sacro-sciatic foramen, divides into two trunks, which are distributed to the subjacent parts.

PLAN OF RELATIONS.

In front.—Peritoneum and ureter.
Outer side.—Psoas magnus muscle.
Behind.—Internal iliac vein, lumbar sacral nerve and psoas muscle.

This artery and the common iliac as regards their length, bear an inverse ratio to each other, the one qeing long when the other is short and *vice.versa.*

The point of division of the internal iliac varies between the upper margin of the sacrum and the upper border of the sacro-sciatic foramen.

The application of a ligature to the internal iliac may be required in cases of aneurism; in wounds affecting one of its branches, or in hemorrhage following amputation of the thigh, &c.

Directions.—Make an incision through the abdominal parietes in the iliac region, in a semilunar direction and to the same extent as for deligation of the common iliac; cautiously divide the transversalis fascia, push the peritoneum inwards from the iliac fossa, and distinguish the external iliac at the bottom of the wound ; trace this artery up until the internal iliac is discovered opposite the sacro-iliac articulation: separate the vein on the *left*, the external illiac on the *right*, and the peritoneum and ureter in *front* of the vessel ; open the sheath, isolate the artery, by passing the *left fore-finger* under it from the *inner side*, and the *right fore-finger* from the outer side, and then hooking it up upon the finger, or grasping it between the thumb and index finger; and, finally pass the ligature around it from *within outwards*.

Remarks.—This operation has been attended with considerable success, but all that was said in regard to the ligature of the common iliac will apply with almost equal force to this deligation of the internal iliac.

One of the cheif dangers is from peritonitis, and the greatest care should be taken not to injure the peritoneum, throughout the various steps of the operation. As soon as it is discovered, the surgeon or an assistant should hold it carefully aside, and its separation continued by gentle touches of the left fore-finger, in the direction of the sacro vertebral articulation, until the vessel is reached. Too much gentleness and caution cannot be exercised in this regard. It is important also not to include the ureter in the ligature, which would prove a most unfortunate mistake. The ureter crosses just at the bifurcation of the iliac artery, and is separated with considerable difficulty from the vessel.—

distinguish it, however, and separate it or abandon the operation so that upon nature and not surgery, may rest the responsibility of a fatal issue.

In making the first incision, great care should be taken not to divide the epigastric artery, or to penetrate the peritoneal cavity, as may be readily done where the muscles are not poorly developed.

Ligature of the External Iliac artery.—This is the chief vessel by which the lower limb is supplied with blood. It passes obliquely downwards and outward from the bifurcation of the common iliac, along the inner border of the psoas muscle, to the femoral arch, where it becomes the femoral artery. The course of this artery is indicated by *a line drawn from the left side of the umbilicus to a point midway between the anterior superior spinous process of the ilium and the symphysis pubes*

PLAN OF RELATIONS.

In front.—Peritoneum, intestines and iliac fascia, spermatic vessels, genito-crural nerve, circumflex ilii vein, symphatic vessel and gland.

Outer side.—Psoas magnus iliac fascia.

Inner side.—External iliac vein of vas deferens and femoral arch.

Behind.—External iliac vein.

Ligation of the external iliac artery may be required for wounds and aneurisms, of the femoral artery, and also for secondary hemmorhage following amputations, when all other means have failed in arresting the flow of blood.

The vessel may be secured in every part of its course save near its upper and lower extremities, the circulation at these points being too rapid to admit of the formation of a sufficiently firm clot to meet the ends in view.

Directions.—Place the patient in a recumbent position; make an incision, commencing an inch above, and to the inner side of the anterior superior spinous process of the ilium, and running downwards and outwards, to the outer end of Poupart's ligament, and from thence parallel with its outer half to a little above the middle: divide the abdominal muscles and cut cautiously through the transversalis fascia; separate the peritoneum carefully from the iliac fossa, and push it towards the pelvis; introduce the index finger, and find the artery pulsating at the bottom of the wound along the inner border of the psoas muscle: separate the iliac vein from the artery, on the inner side, by means of the finger nail; open the sheath, isolate the artery carefully, and pass the ligature under the artery from within outwards, i. e., between the vein and artery, leaving out the small nerve which accompanies the latter.

Remarks.—The direction of the external incision has been much varied by different surgeons. Thus Abernethy cut nearly over the course of the vessel; Sir A. Cooper made the incision from the external margin of the external ring to the anterior superior spinous process of the ilium, following the direction of Poupart's ligament; while Velpeau modified this procedure, without improving on it in the least degree.

The great objection to Abernethy's plan is the anger of subsequent hernial protrusion in consequence of the abdomen being much weakened by the free incisions through its muscular fibres. It has the advantage however of permitting the deligation of the artery at any portion of its course, and of allowing the incision to be extended upwards if necessary' so as to expose the common iliac.

The incision recommended by Cooper is directly across the track of the epigastric and circumflex ilii arteries, as well as the circumflex vein. The spermatic cord is somewhat in the way of this operation. Its chief recommendations are the protection afforded to the peritoneum, and the immunity secured from subsequent hernial protrusions.

The most common evil following these operations, is *gangrene* of the limb, resulting from the curtailment of the sanguinious supply to the part, in consequence of the obliteration of the main channel and the tardy development of circuitous ones. The period at which this mortification occurs is usually about the third or fourth week; and the only means of saving the life of the patient is a speedy resort to amputation.

The greatest possible attention must be bestowed upon the preservation of the peritoneum from all wounds or injury, at every step of the operation. Peritonitis is one of the most serious complications by which the Surgeon can be embarrassed, and the patient's life endangered.

It is important to have the incision as long as practicable, but it must not be carried far enough to implicate the external ring, lest it induce a tendency to hernial protrusion.

Before beginning the operation shave the pubes, and empty the colon by means of an enema.

This operation was first attempted by Abernethy, in 1796, and since that period it has been performed at least 100 times, with a mortality of only 26 per cent. Sir A. Cooper declares that, "this operation may be performed without the least difficulty, and is as easy as tying the femoral artery, there being only one circumstance that occasions the least danger, and that is the epigastric artery which passes up from the iliac vessel, and on the inner side of the incision ; but this however may be avoided."

The distance of the artery from the surface, the great danger of wounding the peritoneum, and its close proximity to important veins and nerves, as well as to the spermatic cord, all go to prove that the deligation of this artery is a more serious and important thing than is supposed by Cooper, and to warn the conscientious Surgeon against an operation into which the mere desire for éclat might possibly hurry him.

The circulation is carried on after the ligation of this artery by means of the gluteal and ischiatic arteries,—the former being the principal one concerned.

It cannot be denied that operations on the iliac vessels generally, are far more successful than upon those vessels above the heart which pertain especially to the trunk, notwithstanding that the former are more deeply seated, surrounded by more delicate structures, and are even of larger calibre.

LIGATION OF THE ARTERIES OF THE SUPERIOR EXTREMITY.—*Ligation of the Axillary Artery.*—The axillary artery commences where the subclavian terminates, at the lower border of the first rib, and becomes the brachial at the lower border of the tendon of the latissimus-dorsi and teres major muscle.

In the normal quiescent position of the limb, the artery forms a gentle curve, the convexity of which is *outwards* and *upwards*.

For convenience of description this artery may be divided into three portions, viz: the portion above the pectoralis major, or *first* part; the portion beneath the pectoralis muscle, or the *second* part; and that portion below the muscle and in the axillary space, the *third* part.

Relations the *first* portion of the axillary artery:

In front.—Pectoralis major, costo-coracoid membrane, cephalic vein.
Outer side.—Brachial plexus,
Inner side.—Axillary vein,
Behind.—First intercostal space and muscle, first serration of serratus magnus, posterior thoracic nerve,

The artery may be tied in this portion, in case of aneurisms or wounds of the *second* portion, but it is not the *point of election*. In some few cases it has been performed with success, but it is always difficult and dangerous.

Directions.—Place the patient on his back, with his shoulders slightly raised, and his elbow a little removed from his body, make an incision three inches long, three quarters of an inch below, and parallel to the clavicle, and terminating at the junction of the deltoid and pectoralis major; cut through the platysma and pectoralis carefully, layer by layer; divide, on a director, the posterior sheath of this muscle which doubles back and has the appearance of an aponeurosis; then bring the arm to the body, and with the end of the director, or the handle of the knife, tear aside the cellular tissue covering the vessel, and carry the finger behind the upper border of the pectoralis minor muscle; draw the vein *inwards* by means of a blunt hook, and pass the needle between it and the artery, from *within outwards*.

Remarks.—This ligature is one of the most difficult to apply, both from the large muscles which have to be cut through, the depth of the vessel, and and the number of of vessels which have to be divided.

It is of the first importance to avoid the cephalic and axillary veins,—the former running along the external border of the pectoralis major, crossing the artery to join the axillary on the inner side of that vessel. The vein is an admirable land mark, and when found should be drawn carefully aside, so that the artery may be reached a little to the inside and behind it.

It is better to tie the subclavian in the third part of its course, for such accidents as seem to demand the application of a ligature above the middle portion of the axillary.

In cases of wound of this artery, the general practice of cutting down upon and tying the vessel above and below the wound should be rigidly adhered to under all circumstances.

RELATIONS OF THE SECOND PORTION OF THE AXILLARY ARTERY.

In front.—Pectoralis major and minor.
Outer side.—Brachial plexus.
Inner side.—Axillary vein.
Behind.—Subscapularis,

The brachial plexus surrounds the artery and separates it from direct contact with the veins and muscles. This vessel is so deeply seated and so completely surrounded by important structures that an operation for its ligation is very seldom attempted. Désault and Delpech have given directions for the proper performance of the operation, but it is now generally condemned because of the facts mentioned above, and the additional consideration of the great depth of the artery and its close investment by important nerves.

RELATIONS OF THE THIRD PORTION OF THE AXILLARY ARTERY.

In front.—Integument, fascia, and pectoralis major muscle.
Outer side.—Coraco-brachialis median nerve, musculo-cutaneous nerve.
Inner side.—Ulnar nerve, internal cutaneous nerve, axillary vein.
Behind—Subscapularis, tendons of latissimus dorsi and teres major, spinal and circumflex nerves.

The artery is usually ligated in this portion, because it is more readily reached and easily isolated.

Directions.—Place the patient upon a bed; separate the arm from the side and supinate the hand; having found the head of the humerus, make an incision over it, through the integuments, about two inches in length, and a little nearer the posterior than the anterior fold of the axilla; carefully dissect through the fascia and areolar tissue, until the median nerve and axillary vein are exposed; displace the former to the *outer*, and the latter to the *inner* side of the arm, bending the elbow so as to relax the muscles; and then, having isolated the artery, pass the needle from the *ulnar* to the *radial* side.

Remarks.—It must be remembered that the axillary artery in about one case in ten gives off a large branch which forms either one of the arteries of the fore arm or a large muscular trunk.

Ligature of this artery is called for in cases of wounds and aneurisms at the upper part of the arm; and, when circumstances admit of its application in the lower portion of the vessel, the operation is simple and easy.

Ligation of the Brachial Artery.—This artery commences at the lower margin of the tendon of the teres major, where the axillary terminates, and extends to about one inch below the bend of the elbow, where it is divided into the radial and ulnar.

The direction of this vessel is marked by a line extending from the outer side of the axillary space to a point midway between the condyles of the humerus, which corresponds with the depression along the inner border of the *coraco-brachialis* and *biceps muscles*. In the upper part of its course, the artery is less internal to the humerus, but below, it is in front of that bone.

RELATIONS OF THE BRACHIAL ARTERY.

In front.—Integument and fascia, bicipital fascia, median basilic vein, median nerve.
Outer side.—Median nerve, coraco-brachialis, biceps.
Inner side.—Internal cutaneous, ulnar and median nerves.
Behind.—Triceps, musculo-spinal nerve, superior profunda artery, coraco-brachialis, bracialis anticus, and bend of the elbow.

The median nerve, at the upper portion of its course is external to it; about the middle of the arm it is in *front* of the artery; and further down towards the elbow, it is upon the *inner side* of the vessel. The basilic vein is at first on the inner side, and then gets in *front* of the artery, and lies in the line of it, for the remainder of its course.

The artery is accompanied by two veins, the venæ comites, which lie within the sheath, in close contact with the main vessel, and are connected together at intervals by transverse communicating branches. At the bend of the elbow, the brachial artery sinks deeply into a triangular space, which contains, also, the radial and ulnar arteries, the median and musculo spiral nerves, and the tendon of the biceps muscle. Occasionally the artery is divided high up the arm, either to unite before reaching the elbow or to be continued to the fore arm as the radial and ulnar arteries.

The artery may be ligatured either in the *upper third* of the arm or in the *middle third* of that member. In the upper portion the *coraco-brachialis* muscle is the *guide* for the operation; while in the lower portion the inner *margin of the biceps* furnishes the proper indication.

Directions for applying the ligature in the *upper portion.*—Place the patient horizontally upon the table, raise the affected limb from the side, and *supinate* the hand ; make an incision two inches in length on the ulnar side of the coraco-brachialis muscle, and divide the fascia carefully as high as the axilla : cut carefully through the cellular tissue and separate the ulna nerve on the inner side, the median on the outer side : open the sheath, and detach the venæ comites which are on either side of the vessel ; and, then, pass the aneurismal needle under the artery from the ulnar to the radial side. The vein is on the *inner* side, and should be carefully avoided. Lisfranc recommends that the position of the median nerve should be found, and that then, placing the four fingers of the left hand, an incision should be made on the inner side of it. Care should be taken in every operation to ascertain whether there are two arteries in the arm, consequent upon a high division of the main trunk, and, in such a contingency, to ligature both of them.

Directions for applying the ligature in the middle of the arm.—Place the patient horizontally upon a table, with the affected limb raised from the side : make an incision along the inner margin of the biceps muscle, two inches and a half in length only including the skin ; open the brachial aponeurosis and carefully carry the basilic vein out of the way : then find the median nerve which is immediately on the edge of the muscle and above the artery, and draw it and the muscle aside, with the blunt hook ; carefully avoid the internal cutaneous nerve on the inner side of the vessel, and open the sheath of the vessel ; then separate the venæ comites isolate the artery, and pass the needle under it from *within outwards*.

The *lower part* of the artery is interesting because of its connexion with the veins usually opened in venesection. The median basilic vein passes immediately in front of the artery, only being separated from it by the fibrous expansion given off from the tendon of the biceps to the fascia covering the flexor muscles. It is important therefore, not to open this vein, if either of the others be large enough to justify an operation, lest the artery be injured by the lancet. Should it become necessary, however, to open it, great care should be observed by the Surgeon, not to wound the artery, &c. If the vein is parallel with the artery, pronate the hand violently, so as to increase the distance between the two vessels, and if the muscles are in the way flex the fore arm slightly, for the same purpose. When the vein is situated immediately over the artery, introduce the lancet horizontally, and compress the artery at the moment of bleeding. Should the artery be punctured, the bleeding may be arrested temporarily, at least, by flexing the fore arm, putting it in a state of pronation, and applying a compress over the wound. It is well also to bandage the whole limb.

Remarks.—As this is the main arterial branch by which the arm, the most useful and exposed of all the members, is supplied with blood, it follows that its deligation, both on account of injury and disease, is a task of very frequent performance. In the battles before Richmond, the number of wounds received in the arm was the subject of universal remark. No accurate statisical information has yet been furnished in regard to this subject, but the author feels assured, from his own personal observation, as well as the assurances of others, that of all the operations performed *upon the field*, at least *half* were for injuries of the superior extremities. The management of the musket and the sabre, the removal of obstructions, &c., necessitate the constant use and exposure of the arms, and thus furnishe an explanation of the fact just mentioned.

Again, the brachial artery is very frequently injured in venesection, both by direct puncture, and development of aneurisms, so as to require the application of the ligature.

The operation may be readily, rapidly, and safely performed, if the anatomical relations of the parts are properly understood, and remembered.

Ligation of the Radial Artery.—The radial artery, judging from its position, is a veritable continuation of the brachial, though it is smaller in size than the ulnar. It commences at the bifurcation of the brachial, an inch below the bend of the elbow, passes along the radial side of the fore arm to the wrist, then runs backwards round the outer side of the carpus, beneath the extensor tendons of the thumb, and runs forward between the two heads of the first dorsal interosseous muscle into the palm of the hand. After reaching the palm it *forms with the deep branch of the ulnar, the deep palmar arch.* It may be therefore divided, for convenience of description, into three parts, viz: *that portion in front of the fore arm; that at the back of the wrist;* and *that in the hand.*

RELATIONS OF THE RADIAL ARTERY.

In front.—Integument, fascia and supinator longus.
Outer side.—Supinator longus, radial nerve, (middle third.)
Inner side.—Pronator radii teres, flexor carpi radialis.
Behind.—Tendon of biceps, &c.

In the upper third of its course, it lies between the pronator radii teres and the supinator longus; and in the lower third, between the tendons of the supinator longus and the flexor carpi radialis.

In the middle third of its course, the radial nerve lies along the *outer side* of the artery: and some filaments of the musculo-cutaneous nerve run along the lower part of the artery as it winds around the wrist. The vessel is accompanied by venæ comites throughout its course.

This artery is tied for wounds and aneurisms.— The *tendon of the flexor carpi radialis* is the the guide for the operation in the middle and lower parts of the arm.

Directions for applying a ligature in the lower third of the fore arm.—Make an incision from half an inch above the wrist joint, two inches in length on the radial side of the tendon of the flexor carpi radialis; divide with another incision the aponeurosis of this tendon; open the sheath and separate the venæ comites; and then isolate and ligate the artery by passing the needle from *without inwards.*

Directions for applying a ligature on the upper third of the fore arm—Make an incision two inches and a half in length, beginning at a point half an inch outside of the middle of the elbow, this should divide the skin only, for fear of injuring the median vein, which ordinarily is on the inner side make another incision, laying bear the supinator longus; raise the internal border of this muscle with the finger or director; then open the sheath isolate, and ligate,—passing the needle from without inwards so as to avoid the nerve.

Directions for applying the ligature on the dorsum of the wrist.—Extend the thumb strongly, so as to cause the abductor longus, and extensor longus pollicis to become prominent; seek for the artery in the depression between these muscles, known as "la tabatiere": separate the thumb from the index finger, and make an incision about an inch long, in the direction of the tendons above referred to; separate the nervous filaments and veins carefully; and then isolate the artery and apply the ligature.

The artery is readily exposed throughout its whole course, but the operation in the upper third is attended with more difficulty than at other portions of the vessel, on account of its greater depth, and the position of the supinator longus muscle.

The operation upon the dorsum of the thumb is fit only for the dissecting room.

It is useless to ligature the radial artery on account of hemorrhage from either the superficial or the deep palmar arch, as the supply of blood from one direction only is thus cut off, leaving a channel equally as broad and deep, communicating with the severed artery. Under such circumstances, as well as for aneurisms and wounds of the hand and fore arm generally, the brachial must be ligatured. In wounds, the general rule must be followed of applying the ligatures at the seat of injury, both above and below the divided surface of the vessel; and when this is impossible, either compression or ligature of the brachial must be substituted.

The origin of the radial varies in the proportion of one in eight cases. Sometimes its point of origin is lower but more frequently higher up.

It is thought by some Surgeons that the ligation of the artery should not be attempted above the middle third, as the operation in the upper third is not only difficult, but calculated seriously to impair the integrity of the muscles.

Ligation of the Ulnar Artery.—This is the larger of the two terminal branches of the brachial. It commences a little below the elbow, then crosses the inner side of the fore arm obliquely to the commencement of its lower half, and runs along the ulnar side of the wrist, until it enters the palm, by crossing the annular ligament, on the outer side of the pisiform bone. After reaching the hand, it forms with the superficialis volæ, a branch of the radial, the superficial palmar arch.

RELATIONS OF THE ULNAR ARTERY.

In front.—Superficial flexor muscles, median nerve, superficial and deep fascia.
Outer side.—Flexor sublimis digitorum.
Inner side.—Flexor carpi ulnaris, ulnar nerve, (lower $\frac{2}{3}$.)
Behind.—Brachialis anticus, profundus digitorum.

At the wrist the ulnar artery is covered by integuments and fascia, and lies upon the anterior angular ligament, with the pisiform bone and ulnar nerve on the *inner side*,—the latter being somewhat behind the vessel.

The *superficial palmar* arch is covered by the palmaris brevis, the palmar fascia, and the integument.

Direction. — The artery is deeply seated in the upper half of the fore arm, beneath the superficial flexor muscles, which in cases of recent wounds, may be divided, but under no other circumstances.

In the *middle and inferior thirds* of the fore arm, this vessel may be secured in this manner: Make an incision on the radial side of the tendon of the flexor carpi ulnaris; divide the deep fascia, and separate the flexor carpi ulnaris from the flexor sublimis; open the sheath, separate the veins, isolate the artery, and pass the needle from the ulnar to the radial side, taking care not to injure the ulnar nerve.

This artery may be deligated in cases of aneurisms, wounds, &c., in either of its main trunks or branches.

It should not be ligatured above the middle third, save in exceptional cases of injury, for fear of permanently injuring the superficial flexor muscle, which must necessarily be cut through in the operation.

In wounds of the palmar arch, it is better to seek for the bleeding orifices, and to ligature each, as compresses produce much irritation and at best are rather paliative than curative measures.

If the hemorrhage cannot be arrested in this way, both the radial and ulnar, or the brachial alone may be tied, which will effectually arrest the flow of blood from the part.

When a compress is used for hemorrhage from the palmar arch, it should be in the shape of a ball —the hand being made to grasp it firmly and the graduated compress applied to the arm; for the purpose of diminishing the amount of blood sent to the part.

LIGATURE OF THE ARTERIES OF THE INFERIOR EXTREMITY.—*Ligation of the Femoral Artery.*—The femoral is a continuation of the external iliac, and extends from Poupart's ligament to the middle of the lower third of the thigh, where it becomes the popliteal. It commences at a point midway between the anterior superior spine of the ilium, and the syph. pubes, passes down the inner aspect of the thigh, and penetrates the adductor magnus muscle. A line drawn from the point just referred to, i. e., midway between the anterior superior spine of the ilium, and the syph. pubes, to the inner side of the internal condyle of the femur corresponds with the direction of the artery, and is nearly above and parallel to it.

In the *upper part of the thigh*, the artery is very superficial, and lies in "*Scarper's triangle*." This triangle is bounded thus : *externally* by the *sartorius muscle*, *internally*, by the *adductor longus*, and *above* by *Poupart's Ligament*, which is its base, its apex being downwards. This triangle corresponds to the depression seen immediately below the fold of the groin, and is nearly equally divided by the femoral artery and vein which run from base to apex.

In this space the artery is crossed in front by the crural branch of the genito crural nerve, and behind by the branch to the pectineus from the anterior crural nerve ; while the anterior crural nerve lies about half an inch to the outer side, imbedded between the iliacus and psoas muscles. The vein, which is included in the sheath with the artery, is on the *inner side*, the vessels being separated from each other by a thin fibrous partition.

In the *middle third* of the thigh, the artery is less superficial, being covered by the integuments and fascia, and overlapped by the sartorious muscle.— It is also enveloped in an aponeurotic canal formed by a dense band which extends from the vastus internus muscle to the tendons of the adductor longus and magnus.

The femoral vein passes beneath the artery, and lies upon its *outer* side; and still more *externally*, is the the long *saphenous nerve*, but not included in the same sheath.

Ligatures are frequently applied to the femoral artery, principally for aneurisms and wounds, and the vessel may be deligated at any point in its course. The operation is however much more difficult in the middle third of the thigh than in the upper part of the course of the artery, because of its greater depth, and the thickness of its aponeurotic covering.

The artery may be tied:
1. Above the origin of the profunda.
2. In the triangle of Scarpa, just above the point

where the artery is crossed by the sartorious muscle.

3 Under the sartorious, just below the apex of the triangle, where the artery is only slightly overlapped by the muscle.

4 Under the sartorius, in the middle part of the thigh.

5 At the outer side of the sartorious, below the middle of the thigh, when the vessel is lodged in the sheath formed by the adductor magnus muscle.

Of these various points, the one *just below the apex of the triangle, where the artery is slightly overlapped by the muscle*, presents the fewest difficulties, and the greatest advantages. This point is about 4½ inches from Poupart's ligament, and is sufficiently below the origin of the profunda to admit of the speedy formation of a firm coagulum within the vessel. The artery can also be readily reached at this point, as it is only covered by the inner edge of the sartorius which can easily be raised, while it serves as a guide to the operator.

Directions.—Place the patient upon his back, with the pelvis slightly elevated; isolate the thigh outwards, and partially flex the limb; follow the course of the artery to the apex of Scarpa's triangle where it ceases to pulsate and is covered by the sartorius. The ligature is to be applied about ¾ of an inch below this point. Make an incision three inches long commencing four fingers' breadth below the fold of the groin, and running directly over the course of the artery; look for the great saphena vein, in the superficial fascia, at the inner side of the incision, and carry it carefully to one side; divide the superficial fascia upon the grooved director; open the cellular tissue beneath with the point of the director, for the whole length of the wound; puncture the fascia-lata, which comes in view, and divide it on the director for about half the extent of the first incision; then draw the inner edge of the sartorius outwards; open the sheath of the artery, isolate the vessel, and pass the aneurism needle carefully lest the vein which is posterior to the artery be wounded.

The application of a ligature to the femoral artery may be required in cases of aneurism or wound of the arteries of the leg, or when hemorrhage of a persistent character follows amputations of the lower extremity.

Larrey tied it above the profunda before amputating at the hip joint, but subsequent experience has demonstrated that this is an unnecessary complication,—increasing materially the difficulty and danger of the operation.

It is a matter of importance not to apply the ligature in the neighborhood of a large branch, lest by so doing, the formation of the blocking coagulum be prevented.

The deligation of the artery within the sheath of the adductor magnus is to be avoided because of the difficulty of reaching the vessel, and the impossibility of preventing the the accumulation of pus within the wound. The point indicated above, is incomparably the best for the operation.

In opening the sheath of the artery, care should be taken to avoid a small nerve which crosses it, and al‑ not to make too large a wound, lest the nutritio˙ of the coats of the vessel be interfered with, a l muscular branches, which are irregular in their origin, divided.

In order to avoid the femoral vein which lies behind and somewhat on the inner side of the artery the needle should be passed from within outwards, the inner side of the sheath being at the same time put upon the stretch. Wounds of this vein are the most serious accidents which associate themselves with this operation, and are usually fatal, producing phlebitis or gangrene.

The ligature of the femoral artery is attended with more success than of any of the large trunks of the body, as is established by the statistics of published cases. In 100 cases collected by Dr. Crisp, only 12 were reported to have died. Secondary hemorrhage and gangrene are perhaps the most frequent accidents which follow this operation, and jeopard its success.

Should secondary hemorrhage occur, four plans of treatment are open to the Surgeon, viz: the employment of pressure; the ligature of the vessel at a higher point; the deligation of the bleeding orifice in the wound; or amputation of the limb. In determining what course to pursue in such a contingency, the Surgeon must follow the light of his own judgment, as no general rules can be established on the subject, and each case prevents features *sui generis* such as furnish the clue to the proper method of treatment.

After the ligature has been applied, the edges of the wound should be brought together with adhesive plaster and stitches, and the limb semi-flexed, somewhat raised, and wrapped in soft flannel or cotton.

The severe pain about the knee which follows this operation, may be relieved by the exhibition of full doses of opium by the mouth, or the administration of morphia subcutaneously.

Ligation of the Popliteal artery.—The popliteal artery extends from the termination of the femoral at the opening in the abductor magnus, to the lower border of the popliteal space, where it divides into the anterior and posterior tibial arteries. This space is lozenge shaped, and is bounded thus: *Externally, above* the joint by the biceps, and *below* the articulation, by the plantaris and the external head of the gastrocnemius. *Internally, above* the joint, by the semi-membranosus, semi-tendinosus, gracilis and sartorius; and *below*, by the inner head of the gastrocnemius. Above it is limited by the apposition of the inner and outer hamstring muscles, and below by the junction of the two heads of the gastrocnemius.

The *artery* is covered *superficially above* by the semi-membranosus: in the middle of its course, by a quantity of fat; and *below* by the margins of the gastrocnemius, plantaris and soleus muscles, the popliteal vein and internal popliteal nerve. The vein is *superficial* and *external* to it until near the termination of its course when it crosses over, and lies on its *inner side*. The nerve is still more *superficial* and *external*, but crosses the artery below the joint, and then, remains upon its *inner* side.

Laterally it is bounded by the muscles which form the confines of the popliteal space.

The operation may be performed in the *upper* or the *lower* part of its course; but in the *middle* of the space, its deligation is attended with much difficulty from the great depth of the artery, and the tension of its lateral boundaries.

Directions for the *upper part* of its course. Place the Patient in the prone position and extend his limbs; make an incision three inches in length through the integument along the posterior border of the semi-membranosus; divide the fascia lata and draw the muscle inwards; find the artery by means of its pulsations; separate the vein, which is on the inner side, and the nerve on the outer side, from the artery, taking care to injure neither the one nor the other; isolate the artery, and pass the needle from *without* inwards.

Directions for Lower portion of its course. Place the patient as before; make an incision through the integument, and in the middle line, commencing opposite the bend of the knee joint, taking care to avoid the saphena vein and nerve; divide the deep fascia on the grooved director, and break up the cellular tissue with its point; separate the vein and nerve from the artery, by drawing the one outwards and the other inwards; isolate the artery and pass the needle from *without* inwards.

Remarks.—Ligature of the Popliteal should only be attempted for wounds of that vessel; but for aneurisms below the joint, it is far better to tie the femoral above. The Popliteal space is so filled with important structures, and the vein, nerve and artery are in such close contact, that some of the best Surgeons, declare it is best not to open the space even in punctured wounds of the Poplite artery.

Operations in this space are also likely to lead burrowing abscesses which may involve the jo and produce the most serious consequences.

Ligation of the Anterior Tibial Artery.—The Anterior Tibial Artery extends from the point at which the Popliteal bifurcates, to the front of the ankle joint where it becomes the Dorsalis Pedis. A line drawn from the inner side of the head of the Fibula to midway between the two malleoli, will be parallel with the course of this artery.

PLAN OF RELATIONS.

In front.—Integument, superficial and deep fascia, tibialis anticus, extensor longus digitorum, extensor proprius pollicis, anterior tibial nerve.

Inner side.—Tibialis anticus, extensor proprius pollicis.

Outer side.—Anterior tibial nerve, extensor longus digitorum, extensor proprius pollicis.

Behind.—Interosseous membrane, tibia, anterior ligament of ankle joint.

In the *upper third* of its course it lies between the tibialis anticus, and extensor longus digitorum; in the *middle third*, between the tibialis anticus and the extensor proprius pollicis; and in the *lower third*, between the tendon of the proprius pollicis and the innermost tendon of the extensor longus digitorum. The anterior tibial nerve lies at first on its *outer* side; then, about the middle of the leg is *superficial* to it; and in the lower part of the limb is again on the *outer* side.

This artery is also accompanied by two veins,

venæ comites, which lie upon either side through out the whole of its course.

The artery may be tied either in the *upper* or the *lower* part.

Directions for the operation in the upper Part.—Place the patient upon his back and extend the limb make an incision about four inches in length midway between the spine of the tibia, and the outer margin of the fibula; divide the fascia and intermuscular septum between the tibialis anticus and extensor communis digitorum, placing the foot so as to relax these muscles and separate them from each other with the finger; having thus exposed the artery, separate the venæ comites on either si , and the nerve on the outer side; isolate the arte and pass the aneurismal needle under it from hout *inwards* so as to avoid the anterior tibial n e.

Directions for the operation in the middle third of the Leg.—Make an incision about three inches in length a ng the external border of the *tibialis anticus* n scle; slit the superficial fascia and aponeurosis r the whole length of the wound and divide them transversely for half an inch or more at each end of the wound, so as to facilitate the separation at the muscles; find the first yellowish intermuscular line which separates the tibialis anticus and the extensor communis digitorum, and open it thoroughly with the finger or the point of the director; flex the foot so as to relax these muscles, and then hold them asunder by means of the finger, or blunt hooks; draw the nerve to one side; then open the sheath, and isolate the artery from its accompanying veins, and pass the needle under the artery.

Directions for ligation of the artery at the lower third just above the ankle joint.—The same general rules will apply. The artery is very superficial and may be readily detected by its pulsations between the tendons of the extensor communis, and extensor pollicis. The nerve is on its outer side, and should be recognized and held carefully aside.

The anterior artery should not be ligatured save for wounds. The *point of election* is the middle third of the limb, as it is more readily reached and isolated at that point. In *the upper* third it is covered by muscles, and cannot be exposed without disturbing them greatly. In the *lower third* of the limb, though the artery is superficial and can be readily found, it is too closely in relation with the sheaths of the tendons, and the ankle joint to justify its ligation save in cases of absolute necessity. The necessity for the double application of the ligature, i. e. above and below the point of division or injury, augments in proportion to the remoteness of the artery from the heart, in consequence of the increased intercommunication by anastomosing branches, which is developed as the vessel recedes from the centre of the circulation.

In isolating the artery advantage will be gained by curving the point of the director. Especial pains should be taken to separate the venæ comites, so as to avoid the induction of phlebitis.

Ligation of the Dorsalis Pedis Artery.—Anatomy. The dorsalis pedis is a continuation of the anterior tibial artery, and extends from the bend of the ankle to the back part of the first interosseous space, where it divides into two branches, the dorsalis hallucis and the communicating.

PLAN OF RELATIONS.

In front.—Integument and fascia, innermost tendon of the extensor brevis digitorum.
Tibula side.—Extensor proprius pollicis.
Tibrila side.—Extensor longus digitorum, anterior tibial nerve.
Behind.—Astragalus, scaphoid, internal cuneiform, and their ligaments, and the anterior tibial nerve.

It is accompanied by venæ comites which lie on its outer side.

Directions.—Make an incision through the integument two inches and a half in length, on the fibula side of the extensor proprius pollicis, in the interval between it and the inner border of the short extensor muscle; divide the fascia and expose the artery; separate it from the venæ comites, and anterior tibial nerve on the outer side, and pass the aneurismal needle beneath it from *within outwards*.

This is a simple operation and may be performed in cases of recent wounds or of hemorrhage following amputations of the toes. Care should be taken not to make the incision farther down than the back part of the first interosseous space as the artery divides at that point. It may be tied at any part of its course, but the middle of the tarsal arch is the point usually selected. Compression may be easily effected by pressing it against the tarsal bones, and this should always be fully tried before resorting to an operation. Occasionally the Dorsalis Pedis is developed into a vessel of large size, but not unfrequently it is almost entirely deficient. When it does not send terminal branches to the toes, they are supplied by branches from the internal plantar artery. Sometimes the place is entirely supplied by a large anterior peroneal artery.

Ligation of the Posterior Tibial artery.—The Posterior Tibial is larger than the anterior and extends from the lower border of the popliteus muscle, to the fossa between the inner ankle and heel, where, beneath the origin of the abductor pollicis, it divides into the internal and external plantar arteries. At its origin it lies opposite the interval between the fibula and tibia; as it descends, it approaches the inner side of the leg, lying behind the tibia; and in the lower part of its course, it is situated midway between the inner malleolus and the tuberosity of the os-calcis.

PLAN OF RELATIONS.

In front.—Tibialis posticus, flexor longus digitorum, tibia, and ankle joint.

Inner side.—Posterior tibial nerve, upper third.
Outer side.—Posterior tibial nerve, lower two thirds.
Behind.—Gastrocnemius soleus, deep fascia and integument.

It is covered by the intermuscular fascia, which separates it above from the gastrocnemius and soleus; while, in the lower third, where it is more superficial it is covered by the integument and fascia, and runs parallel with the inner border of the tendo Achillis. It is accompanied by two veins, and by the posterior tibial nerve which is just on the inner side of the artery, but soon crosses it, and is on its outer side for the greater portion of its course.

At the ankle, the tendons and blood vessels are arranged in the following order: First the tendons of the tibialis posticus and flexor longus digitorum, lying in the same groove, behind the inner malieolus, the former being the more internal. Externally is the posterior tibial artery, having a vein on either side, and still more externally, is the posterior tibial nerve. About half an inch nearer the heel is the tendon of the flexor longus pollicis.

Directions for the application of a ligature in the upper third.—Half flex the leg and lay it upon the inner side; make an incision four inches in extent beginning at a point ¾ to 1 inch behind the inner edge of the tibia, and running parallel with that bone; divide the superficial fascia and aponeurosis, to the same extent, taking care to avoid the saphena vein, which runs up nearly in the direction of the cut; make an incision across the aponeurosis at the two extremities of the wound; separate the cellular connexions of the internal head of the gastrocnemius, on the anterior surface, with the fore finger or director, and draw the muscle aside with the blunt hook; divide the belly of the soleus layer by layer in the direction of the external wound, and at the distance of ¾ of an inch from the tibia; cut the tendonous fibres of this muscle on the grooved director, for the whole length of the original incision; then divide the deep seated aponeurosis, cautiously and in the same manner;

open the sheath of the artery, isolate the vessel, and pass the needle below, from within outwards.

Directions for the application of a ligature at the middle third of the leg.—Place the Patient as before : make an incision three inches long obliquely downwards and backwards from the posterior angle of the tibia to the inner border of the tendo Achillis, so as to cross diagonally over the intermuscular groove in which are lodged the vessels ; divide the superficial fascia and aponeurosis in the same direction ; glide the forefinger into the bottom of the wound, and under the tendo Achillis, so as to detach its cellular connexions freely; draw the belly of the soleus, which now comes in view upwards and backwards, or divide it if necessary; puncture the deep seated aponeurosis, insert the director and divide carefully ; then open the sheath of the vessel, isolate, and tie the artery.

Directions for the application of a ligature to the Posterior Tibial Artery at the ankle joint.—Place the limb as before; make a similunar incision through the integument, two inches and a half in length, midway between the heel and the inner ankle; divide the subcutaneous cellular membrane, and then cut through the *internal annular* ligament, cautiously upon the grooved director; open the sheath of the vessel, separate it from the venæ comites, isolate, and pass the needle from the heel towards the ankle in order to avoid the posterior tibial nerve, care being taken not to include the veins.

Directions for the application of a ligature to the Posterior Tibial artery in the lower third of the leg.—Place the Patient as before; make an incision about three inches in length, parallel with the inner margin of the tendo Achillis: carefully avoid the internal saphena vein, and divide the two layers of fascia upon a grooved director; open the sheath, separate the artery from the venæ comites, isolate, and introduce the needle so as to avoid the nerve which is on the external side.

The depth of the artery in the upper and middle thirds renders it very difficult to tie the vessel at these points, and it is only justifiable in cases of wounds of the vessel.

In aneurismeal tumours of the middle third, it is better to ligate the femoral, rather than to operate in these localities.

When the sole of the foot is wounded or when obstinate hemorrhage follows amputation of the toes, &c., the artery should be tied either at the ankle joint or in the lower third of its course.

The latter steps of all these operations may be much facilitated, by flexing the leg upon the thigh and extending the foot so as to relax the muscles.

The incision must be made from above downwards when the right leg is operated on, and from below upwards where the ligature is applied to the left limb.

Guthrie recommended and practised ligation of the popliteal artery in cases of wounds complicated with extensive effusion of blood between the muscle; but it would be far better to tie the femoral under such circumstances.

When this artery is tied for wounds, no regular operation can be performed, but an incision of sufficient length should be made through the gastrocnemius and soleus, taking the wound for its centre. Two ligatures must invariably be applied under these circumstances, the one above and the other below the point of division, so as to prevent the possibility of hemorrhage either from the cardiac or the distal side of the vessel.

In wounds of the foot, compression should be made upon the artery, at a point about a finger's breadth behind the inner malleolus, before resorting to an operation. Pressure upon this point soon becomes very painful, and should not be persisted in.

Ligation of the Peroneal Artery.—The peroneal artery rises from the posterior tibial, about an inch below the popliteus muscle, and terminates upon the outer side of the os-calcis. It rests first upon the tibialis posticus, and for the greater part of its course in the fibres of the flexor longus pollicis, in a groove between the interosseous ligament and the bone. It is covered in the upper part of its course by the *soleus* and deep *fascia*; and below by the flexor longus pollicis.

PLAN OF RELATIONS.

In front.—Tibialis posticus, flexor longus pollicis.
Outer side.—Fibula.
Behind.—Soleus, deep fascia, flexor longus pollicis.

This artery rarely requires to be tied, except in cases of compound fracture, or punctured wounds, when no general rules can be followed. It is too deeply seated above, and too small below for an operation, so that it is only in its middle portion that a ligature is applied. This artery lies between the tendo Achillis and the fibula, while the posterior tibial is on the opposite side, between the tendo achillis and the internal malleolus.

For statistical information in regard to the ligation of arteries in the city of Richmond, refer to table "H" of appendix.

CHAPTER VII.

DISLOCATIONS.

Lawrence defines dislocation to be "a permanent separation of one, two, or more bones that are naturally articul ed—a separation that is generally produced by external violence." According to this definition every bone in the body is liable to this accident, yet many of them are so firmly attached as to preclude the possibility of such a result save by the employment of an amount of force which produces other effects of so much graver character as to render their mere separation a matter of subordinate consideration. The bones which compose the skull, for instance, hardly admit of being detached the one from the other, save by a degree of violence which produces the most serious consequences to themselves and the subjacent parts. The same remark applies to bones of the pelvis, and, in fact, to all bones, connected by plain surfaces almost as broad as themselves, such as the vertebræ, the tarsus and the carpus.

Far the greater number of these accidents occur at those articulations which are known as the ginglymoid or hinge like joints, and the orbicular or ball and socket joints.‡ The former are neither so firmly held together by ligaments nor so strongly supported by muscles, and hence, their separation is a matter of easy accomplishment.

The orbicular, for the same reason, require less force to separate them than the ginglymoid; thus dislocations occur with more ease and frequency at the shoulder than at the elbow, at the hip than at the knee, and so on for other similar articulations.

VARIETIES OF DISLOCATION.—Dislocations may be complete, incomplete, spontaneous, simple, compound, complicated, congenital, recent, ancient primitive or consecutive.

Complete Dislocation.—When the articular surfaces are entirely separated, the dislocation is said to be *complete*.

Incomplete Dislocation.—When the bones are only partially separated, the dislocation is said to be *incomplete*. Practically, there is not a great deal known concerning this variety of dislocation at the *orbicular joints;* but instances have occurred where the head of the humerus was found on the edge of the glenoid cavity. In the hinge like joints, as the knee, elbow, and ankle, the osseous surfaces commonly remain partially in contact.

Spontaneous Dislocation.—This occurs in consequence of disease. When the ligaments which

---

‡ "Much more depends upon the relative exposure of the joint," remarks Hamilton, "than upon its anatomical structure."

connect the bones are altered by disease of the joint, one of the bones may be thrown out of position by the action of the muscles, the ordinary checks and balances being removed—an occurrence which not unfrequently takes place at the hip joint, and is occasionally seen in the knee. Sometimes, in children, there seems to be an entire relaxation both of the muscles and ligaments surrounding the shoulder joint, and spontaneous dislocation occurs, the limb falling from its normal position, by the force of gravity alone. This is a grave accident, requiring time, patience, and skill to secure a permanent retention of the parts in their natural position.

Simple Dislocation.—This dislocation is called *simple* when unattended by fracture of the bone, laceration of muscular tissue, injury to nerves, division of blood vessels, &c.

Compound Dislocation.—A compound dislocation is one in which there is an external wound connecting with the separated parts. The skin is usually made tense by the presence of a portion of the bone in an abnormal position, and in some instances it is ruptured, making an external wound, through which the osseous structures protrude or not, according to the circumstances of the case. When this rupture occurs a compound dislocation is the result.

Complicated Dislocation.—When in conjunction with the separation of the bones, there occurs fracture of the articulating surfaces, muscular laceration, injury of important nerves, division of large arteries, &c., the dislocation is said to be complicated·

Congenital Dislocation.—When from malformation of the articulation the bones cannot remain in contact, the dislocation is styled congenital.

Recent Dislocation.—A luxation which has taken place within a period of a few days or at least a few weeks is styled "*recent.*"

Ancient Dislocation.—A luxation which has existed for a longer period is considered an "ancient dislocation," though the exact point of time at which it ceases to be "recent" and becomes "ancient" has not been fully determined.

Primitive Dislocation.—When the bone remains nearly or precisely in the position into which it has been first thrown by the force brought to bear upon it, the luxation is "primitive."

Consecutive Dislocation.—When the original position of the bone has been changed, in consequence of muscular action, attempts at reduction, or from any other cause, the luxation is called "consecutive." Thus a "primitive" dislocation upon the ischiatic notch may become a "consecutive" dislocation upon the dorsum-ilii or *vice versa*.

CAUSES OF DISLOCATION.‡—The causes which operate in the production of dislocations may be divided into *immediate* and *remote*.

---

‡ Malguigne after an analysis of six hundred and forty three cases of dislocation, states that "hexations are very rare in infancy, and that the frequency increases gradually up to the fifteenth year—then more rapidly up to the sixty fifth year, from which period onward they become more rare." The deduction from this statement is that age, as a predisposing cause, is most active in middle life, less active in advanced life and least active in early life.

Immediate causes are those agencies which exercise a direct instrumentality in separating the articulated bones. Under this head are comprised external violence, muscular contraction, and a combination of the two.

External violence.—This may act either *directly* by pulling or twisting the parts asunder—as when the foot is displaced by a turn of the ankle, when the thumb is dislocated backward by a blow, or when the arm is torn from its socket by machinery—or *indirectly* when the force acts at a distance from the joint and the bone is thrown from its socket by the "lever like movement of the shaft"— as takes place when the head of the humerus is dislocated by a fall.

Muscular action.—Muscular action may cause the displacement of a bone even when the parts are in a healthy condition. Thus the lower jaw may be dislocated by excessive gaping, and the humerus driven from its place by making a violent muscular effort as in throwing a stone, striking a blow, &c. When the joint has been weakened by previous disease, dislocation readily results from muscular action as can be easily understood.

Combination of external violence and muscular action.—That dislocation may be occasioned by the combined influence of these two causes, when neither would be sufficient of itself to produce such a result, is evident. The usual manner, however, in which these two agencis act together is conjointly but not contemporaneously. Thus, in dislocation at the orbicular joint, after the head of the bone has been thrown out of the cavity by external violence, it is still farther displaced by the action of the muscles which surround the part.

Remote causes are those influences which, by relaxing the ligaments, weakening the muscles, altering the articular surfaces, &c., &c., predispose the parts to separate, and facilitate the action of the various agencies described in detail in the preceding paragraph. An abundant secretion of synovia, even when no organic change has taken place in connexion with the articulation, belongs properly to the catagory now under consideration.

SYMPTOMS OF DISLOCATION.—The symptoms or signs by which dislocations may be recognised are; pain;‡ loss of symmetry; change in the direction of the limb; alteration in the length of the member; preternatural immobility; swelling of the surrounding parts; and loss of normal function.

It may be distinguished from fracture by the absence of crepitus; by the fixedness of the member; and by the failure of the bones to separate after having been properly approximated. Notwithstanding, that these three signs constitute the usual distinction between dislocation and fracture, it is impossible to rely exclusively upon any one of them in determining the diagnosis. Each may in turn associate itself with either accident, and it is only by considering them together as whole, in conjunction with other circumstances, that a correct opinion may be formed in a multitude of cases.

---

‡ The pain of dislocation is more intense than that of fractures in consequence of the pressure of the ends of the bone upon the nerves

TREATMENT OF DISLOCATION.—The general treatment of dislocation consists in:

The *reduction or return* of the bones to their *normal relations*.

The *retention* of the bones in their *original position*.

Reduction.—In returning the bones to their original relations, the Surgeon has four great obstacles to contend with and to overcome, viz:

Muscular contraction; the anatomical construction of the joint; the smallness of the tear in the capsule and the difficulty of finding its direction and position—this is especially true of the hip joint;—and the development of ligamentous bands forming new but powerful attachments between the head of the bone and the surrounding parts.

The *first* obstacle is to be overcome by means of what is known as manipulation;‡ and by extension and counter extension;—aided by the administration of Chloroform, by the exhibition of nauseants and depressants; by bleeding; by the warm-bath; by the subcutaneous introduction of opium or some one of its preparations, particularly the salts of morphia. In regard to the latter mode of reaching and relaxing muscular fibre, the author would state, that after much experience and many carefully conducted experiments, *he is* so thoroughly convinced of the great value of this practice generally as to induce him to recommend it in the

---

‡ This is familiarly known as Reid's method, though it dates as far back as Hippocrates, and was successfully practised by Wiseman in 1676, for certain luxations at the hip joint. So far as the United States are concerned, to Physic and Nathan Smith the credit is due of introducing this method of treating dislocations. Reid did not make his report until the year 1851, some forty years after the successful experiment of the above named Surgeons.

most unqualified terms, to the profession. Under the head of extension and counter extension are included the various mechanical contrivances which are employed for the purpose of overcoming muscular contraction and of returning the bones to their original position.

When manipulation has failed, extension and counter extension may be made with the hands of the Surgeon or his assistants, with the compound Pulleys, with the simple rope Windlass, with Jarvis' adjuster, and with such other similar appliances as may suggest themselves in this connexion. In this way we are enabled to exert much more power and to overcome the contraction of the muscles by steady and gradual resistance, but there is always danger of doing serious injury to the soft parts; and hence, the importance of using no more force than is absolutely necessary and of proceeding with great caution and circumspection.

When individual dislocations are considered, the proper directions for using of these various mechanical contrivances will be explained in detail.

The general rules for the application of extension may be thus summed up.

1. Protect the skin by means of a wet roller before applying any powerful extending force.

2. Apply the force slowly, gently and continuously, carefully avoiding any jerking of the parts, lest the artery be severed, the muscles excited to still stronger contractions, &c.

3. The traction should be made in the axis which the limb has acquired by its change of position

without reference to its normal direction or the situation of the articulation.

4. In dislocations at the hip joint, apply the extending force to the femur,—the bone displaced; but in dislocations at the shoulder joint, apply the extending force to the fore-arm, using the whole limb as a lever.

5. Do not employ the Pulleys, the adjuster, &c., until an effort at reduction has been attempted by making traction with the hands, &c., aided by Chloroform and such other agents as tend to relax the muscles.

The author has succeeded in relaxing muscular contraction of an obstinate and decided character by the subcutaneous injection of morphia immediately over the track of those muscles offering most resistence to the return of the bone and he therefore recommends this procedure as an invaluable adjuvant in the accomplishment of the indication in question, particularly if there be but little tumefaction about the parts, and only a slight development of adipose tissue.

The *second* obstacle is to be overcome by obtaining an exact knowledge of the anatomical structure of the joint, and using such mechanical appliances as may be necessary to tilt or lift the head of the bone over any projecting and opposing eminence into its proper cavity. Some times this is effected by using the limb as a lever of various degrees, and then again by the direct application of force in such away as to raise the limb bodily over the obstruction, relying upon muscular contraction to carry it into its normal position. This subject will also be more particularly dwelt upon under the next section of this work.

The *third* obstacle is to be met and disposed of by endeavouring to find the particular locality of the *tear*, and ascertaining the direction and position in which the dislocated head corresponds to the hole in the capsule. The dislocated head does not always preserve the original position in which the luxating force places it, but by means of an abduction, flexion, &c., which subsequently follows, is forced into a new position. *All attempts at reduction must therefore be commenced by restoring the dislocated bone to its primitive position, and causing i to glide from that into its normal situation.* These observations, of course, apply only to ball and socket joint, and have a particular reference to dislocations of the hip.

The *fourth* difficulty is to be surmounted by operating before the new attachments have formed, or rather before they have become thoroughly organized.

The period beyond which reduction should not be attempted varies according to the nature of the dislocation and the concomitant circumstances. It may be practised at a much later day in luxations of the orbicular than of the ginglymoid joints and this remark applies particularly to those at the shoulder, as all experience demonstrates. Sir A. Cooper declares, emphatically, that "the latest period at which reduction can be safely effected, even in this dislocation, does not exceed *three months;* while for the hip *eight weeks* is the proper limit." It is undoutedly true, that these dislocations have been reduced, with entire safety t the patients; but these are the exceptions rather than the rule, and should be so regarded by the Surgeon.

Retention.—When the bone has been returned to its normal situation, it must be retained there by proper splints and bandages, and rest enjoined, at least for several days.

If symptoms of inflammation show themselves, they should be arrested by the prompt and persistent application of cold water.

In dislocation, it should be remembered, that the principal indications as well as the chief difficulty consists in *reduction*, that is, restoring the parts to their natural *status*; while on the other hand, in fracture, the most important desideratum is to employ means to *retain* the parts in apposition, after they have been reduced.

The above rules hold good for the treatment of simple dislocation.

Treatment of Compound Dislocations.—These are very serious injuries, owing to the peculiar susceptibility of the parts which enter into the formation of joints, to take an inflammatory action. There is usually little or no difficulty in reducing the dislocation, or in retaining the bones in position; but the great danger is in the subsequent inflammation, suppuration, &c., which are likely to ensue. If there be a reasonable probability of securing union by the *first* intention, the parts should be brought together and cold water dressings employed; but on the other hand, if the joint be large, and there s much laceration of the soft parts, the limb should be amputated.

Wounds of this character are more favorable when occurring in their upper than the lower extremity for reasons already given at length in another portion of this work.

Treatment of Complicated Dislocations.—Should be treated on the same principles, precisely as the ast. Indeed the two so frequently occur contemporaneously that it is unnecessary to establish different rules for their management. If the bony parts immediately involved are fractured, resection may be successfully practised unless the soft parts are too much injured, when amputation must be speedily resorted to, as the only means of preserving life.

When *fracture* of the *shaft* of a bone is complicated with dislocation of its head, great difficulties will necessarily present themselves in the way of a proper reduction. It is much safer to reduce the dislocation without waiting for the bone to unite as the period required for this process, would carry the surgeon far beyond the time when reduction is esteemed a practicable measure. The fractured limb must be put up very carefully in wooden splints, before extension is made.

Particular Dislocations.—*Dislocation of the lower Jaw.*—The inferior maxillary bone may be either completely or *partially* dislocated. When completely dislocated, both condyles slip beyond the eminentiæ articularis into the zygomatic fossa, while the coronoid process hitches against the malar bone and the axis of the same is directed obliquely forwards. When the bone is partially dislocated, one condyle remains in position while the other is carried forwards into the zygomatic fossa.

A sub-luxation is also described by Sir Ashly Cooper, which is most frequently met with in young and delicate women, in which the head of the bone appears to slip before the internal articular cartilage, so as to prevent the closure of of the mouth.

Causes.—Sometimes this dislocation is caused by direct violence,—as by blows, kicks, falls, &c. Again, in gaping, yawning and laughing, the muscles are put too violently upon the stretch and the condyle is carried beyond the glenoid cavity. The jaw has also been dislocated in attempts made to draw teeth, by a sudden action of the hand, depressing the chin to too great an extent.

An imperfect dislocation of the jaw is sometimes occasioned by a relaxation of the ligaments surrounding the joint.

Symptoms.—In partial dislocation the mouth is not so widely open as in complete dislocation, but the patient cannot close it in consequence of the condyloid process being carried against the zygoma. The chin is carried to the opposite side; the incisor teeth are advanced upon the upper jaw; saliva is somewhat increased in quantity; and articulation is difficult.

In complete dislocation the mouth is widely opened and cannot be closed; deglutition and speech are much impaired; the chin is lengthend; the saliva dribbles over the lips in consequence of pressure on the parotid glands; the cheeks are flattened;

the lower line of teeth are advanced beyond the upper; and there is a depression in front of the meatus, and a prominence in the temporal fossa between the eye and the ear.

Treatment.—Stand before the patient and apply the thumbs well protected to the molar teeth on either side; and then depress the angle of the jaw forcibly, and at the same time raise the chin by means of the fingers passed under it. When only one side is luxated, the efforts at reduction should be confined to that side alone In *subluxation*, constitutional remedies, such as iron, valerian, &c., should be administered, and repeated blisters applied directly over the joint.

Should the ordinary means fail of their object, the following plan may be resorted to : Place some hard substance, as the handle of a spatula, a piece of wood or ivory, between the molar teeth or the upper and lower gum, on either side, or transversely from one to the other; step behind the patient, and pass the hands forward under the chin ; push the chin up forcibly, so that by means of the wood between the teeth, as a *fulcrum*, and the bone itself as a *lever*, the head may be prized out of its new socket, and carried by the muscles over the eminentia articularis into the glenoid cavity.

The four-tailed bandage may then be applied, and the patient made to refrain from talking, eating solid food, laughing, &c. Very old dislocations may be reduced by the process last described.

Dislocations of the clavicle.—The clavicle may be dislocated at either of its extremities, that is, at its sternal or acromial end, but this accident is rare compared with that of fracture of the bone, because of the strength of its ligamentous attachments.

The sternal end may be luxated either *forwards, backwards,* or *upwards,* being thrown *before, behind,* or *above* the sternum.

The acromial end may be dislocated and placed upon the *upper surface of the acromion,* upon the *anterior part of the spine of the scapula, under the acromion* and *beneath* the coracoid process.

Symptoms.—As the clavicle is very superficial, the changes of conformation which accompanies these various dislocations are so obvious as to render a recognition of the accident a matter of great facility. The head of the bone can be distinctly felt in each of the dislocations referr to, making its diagnosis easy and certain.

Causes.—External violence,—as blows, kick &c.

Treatment.—Reduction is easy but retention difficult, because the accident cannot occur without the rupture of the strong ligaments which ordinarily hold it in position.

Treatment of dislocations of the sternal end.—The dislocation forwards is to be reduced by pushing the shoulder outwards and bending it *backwards,* and the parts retained in position by means of a pad and a figure of 8 bandage applied firmly over the displaced end of the bone—strips of adhesive plaster may be substituted with advantage for the latter.

The dislocation upwards is of extremely rare occurrence, but when ascertained, should be treated by means of a bandage and pad, together with the elevation of the elbow.

The dislocation backwards is not of common occurrence, though there are quite a number of cases on record. It generally results from the point of the shoulder having been driven *upwards;* or by the hand being drawn violently *forwards;* or by the direct pressure of the clavicle *backwards.* The treatment consists in making a fulcrum of the fist or knee in the axilla, and then bringing the elbow well to the side. In this way the dislocation is *reduced* with facility. *Retention* is difficult, and must be accomplished by the figure of 8 bandage tightly applied to the shoulders, and crossed over a large pad placed in the middle of the back, the elbow being at the same time fixed to the side. Adhesive straps may be substituted for the ordinary bandage as they adhere to the skin and remain much more permanently in position.

Treatment of luxations of the Acromial end.—The dislocation of the head of the bone upon the *upper* surface of the acromion can be recognized and reduced easily by manipulation. The shoulder should be pushed *upwards, outwards,* and *backwards,* and held in that position by the same means as those employed for fracture of the clavicle,—all of which will be fully described under the head of fractures, &c. Adhesive straps passed from the shoulder to the elbow, embracing the arm, are admissible substitutes for other and more complicated arrangements. It may be well also to place a pad in the axilla and to bind the arm to the side.

Dislocation under the Acromion.—Nélaton states

that there are only three cases of this luxation on record. It certainly is of very rare occurrence. The treatment is precisely the same as for fracture of the clavicle.

Dislocation beneath the coracoid process simply requires the clavicular bandage.

Dislocations at the Shoulder Joint.—The humerus may be dislocated in four directions, viz: downwards in the axilla; forwards under the clavicle; backwards upon the scapula.

Dislocation downwards.—This dislocation is of most frequent occurrence.

Causes.—Falls upon the top of the shoulder; blows upon the shoulder; violent abduction of the arm; &c.

Symptoms.—The Acromion projects; the rotundity of the shoulder is lost; a round body can be felt in the axilla; the arm is lengthened, numbed, and carried out from the body three or four inches; the hand cannot be placed upon the opposite shoulder while the elbow touches the thorax; there is great pain when the elbow is forced against the side.

Treatment.—Reduction is accomplished either by manipulation or by the employment of force.

Manipulation.—Administer Chloroform; carry the elbow about 45° from the side; flex the forearm at a right angle with the arm so that the palm of the hand presents to the patient's abdomen; then, using the forearm as a lever, rotate the head of the humerus forwards and upwards by making the hand describe a semi-circle from before backwards until the palm of the hand looks up, the elbow being kept off from the side; holding the forearm in its semi-flexed position, with the palm of the hand looking to the operator, carry the elbow gently into the side; then quickly rotate the head backward and upwards by reversing the motion of the forearm so as to cause the hand to describe an entire circle.

In the anterior and posterior dislocations carry the arm as nearly perpendicularly upwards as possible, or in such a position as will throw the head of the bone into the axilla, and then proceed as before.

During the operation the scapula should be firmly fixed and firmly held by reliable assistants.

Employment of force.—The dislocation may be reduced, when manipulation has failed, by means of the heel in the axilla; by means of the knee; by means of Pulleys; by Jarvis' adjuster, &c.

By the heel placed in the axilla.—This is the oldest and most convenient process, and will answer for a majority of recent dislocations.

Directions.—Place the patient upon his back; administer Chloroform or Ether freely; seat yourself along side, and place the foot in the axilla; take hold of the wrist, and fix one foot firmly on the ground; then draw the limb steadily downwards; and when the head of the humerus is disengaged, and drawn out of its new bed, carry the hand across the patient's body, employing the foot as a fulcrum to turn the bone into its proper situation. Additional force may be employed by fastening a bandage around the arm and carrying it over the shoulders of the Surgeon, so that the weight of the body may be used also as an extending force.

If this be not sufficient, still greater power may be gained by passing a towel under the axilla, and making an assistant pull upwards and backwards while the extending force is applied as just described.

Process with the knee.—This is precisely the same in principle as the last.

Directions.—Seat the patient in the chair; take a stand by his side, rest one foot upon the chair, and place the knee in the axilla; then seize the arm about the elbow with the right hand; steadying the acromion with the left, and draw the limb forcibly downwards; and, when the head has been disengaged, carry the arm inwards across the patient's body.

Process by the Pulleys.—If the muscles contract vigorously, or the dislocation be of long standing, so that it does not yield to the various processes described above, it may become necessary to use still additional force and the Pulleys may be employed.

Directions.—Place the patient in a firm chair; fold a table cloth or sheet to the breadth of eight or ten inches, and place it around the chest so that its middle portion is applied to the axilla, and attach its ends to some fixed point in the floor or wall; pass a wet roller round the arm just above the elbow, and upon this fasten either a strong worsted tape, by means of a *clove-hitch*, or a towel properly adjusted so as to excoriate as little as possible; and to this hitch a towel, apply the extending force, and make firm but steady traction. While this is being done by assistants, stand on the outside of the arm. keep it bent, and rotate the humerus on its own axis as much as possible. Sometimes by placing the knee in the arm pit, the reduction will be much facilitated.

The treatment after reduction is simple. Brace the arm by the side of the body, either by long strips of adhesive plaster, or the roller bandage and support the forearm and hand in a sling. Continue this until the *tear* in the capsular ligament has united, and the muscular tissues have returned to their normal condition of quiescence.

Compound and complicated dislocations should be treated upon the principles already established in the section which treats of dislocation in general.

Dislocation forward under the clavicle.—*Causes.* The causes are the same as for the last dislocation, except that the direction of the impulse slightly varies. In many instances this is consecutive upon a dislocation into the axilla.

Symptoms.—There is a depression under the outer end of the acromion; the elbow is separated from the body and carried a little backward; the axis of the arm is thrown inwards towards the middle of the clavicle; the head of the bone may be felt under the clavicle; the hand cannot be placed upon the opposite shoulder while the elbow remains in contact with the chest; and there is pain or numbness.

Treatment.—The treatment is the same as for the last dislocation, save that the extension has to be made at first somewhat in a line backwards from the body until the head of the bone has escaped beneath the coracoid process; the extension must be made downwards and outwards. Subsequently pull downwards or even upwards, and press the head of the bone into its socket. Retain as before.

Dislocation backwards upon the scapula. This form of dislocation is seldom met with.

Causes.—Falls and muscular exertion, with the arm in a position exactly the reverse of the last.

Symptoms.—There is a projection under the spine of the scapula; and a corresponding depression under the acromion; there is a wide space between the head of the bone and the coracoid process; the ax is of the shaft is directed upwards and outwards the arm is in contact with the body and carried across the chest ; the humerus is rotated inwards; and the hand cannot be placed upon the opposite shoulder.

Treatment.—Sir Astley Cooper recommends the same plan of treatment with pulleys, &c., as in the downward dislocation, and that the extension should be made downwards and outwards. Vidal de Cassis insists that extension shall be made in the direction in which the limb is found; and in this he is sustained by a majority of those who have had the accident to manage. Try either plan, or both in turn, but take especial care to fix the scapula. The bone is retained in place by placing a compress against the head of the humerus and beneath the spine of the scapula, and retaining them in position by means of a roller bandage.

There are other partial dislocations of the humerus, for an account of which the reader is referred to standard works on the subject.

Dislocations at the Elbow Joint.—Numerous luxations occur at this joint, viz : dislocation of the ra dius and ulna backwards; dislocation of radius and ulna forwards; dislocation of both bones laterally; dislocation of the ulna backwards; dislocation of radius forwards; dislocation of radius backwards; and dislocation of radius outwards.

Dislocation of Ulna and Radius backwards.— This accident is plainly marked by the change in the form of the joint, and by its great loss of motion. There is a considerable projection posteriorly; on each side of the olecranon there is a depression; the articulating end of the humerus can be felt in front; the hand and fore arm are in a state of supination, and cannot be pronated; and the fore arm is slightly flexed on the arm. The coronoid process is frequently broken, and if so, may be felt loose in front of the joint; but if not, it will be found fixed against the posterior surface of the humerus.

Treatment.—This dislocation may be reduced thus: seat the patient; take hold of his wrist, and place your knee on the inner side of the elbow joint; bend the fore arm and press upon the radius and ulna firmly with the knee, so as to separate them from the humerus, and to remove the coronoid process from the posterior fossa of that bone; while this is being done, gradually flex the fore arm, and the bones will slip into their respective sockets.

Apply a bandage, keep the arm in a flexed position, use cold lotions, and support the limb with a sling. When the coronoid process is broken, keep it firmly in its place by means of a compress and adhesive straps.

This accident is usually caused by an attempt to catch, on the imperfectly extended arm, while falling.

Dislocation of both Bones forward.—It is almost impossible for this accident to occur without a fracture of the olecranon process: though it may do so in rare cases. It may be recognized by the *elongation* of the *fore arm*; the projection of the condyles of the humerus; the depression of the posterior surface of that bone; and, when the olecranon is broken off by the presence of that process behind, and the great mobility of the fore arm.

Treatment.—The same process is to be followed as in the last case, only the force used must be greater. Put the arm up in firm angular splints, keep the hand semi-pronated, apply cold lotions, and use the sling.

Dislocation of both Bones laterally.—This dislocation may occur on either side, but generally is an incomplete one, either the radius hitches against the internal condyle, or the ulna against the external, and prevents an entire separation of the articular surfaces. This may be recognized by the peculiar deformity; loss of motion; the movements of the radius under the hand when the arm is rotated; the position of the ulnar either on the inner or outer condyle; the radius forming a protuberance behind, and on the outer side of the humerus; by the great projection of the condyle; and by the hollow above the olecranon.

Treatment.—Reduce the dislocation by bending the arm powerfully over the knee and making traction at the wrist. As soon as the radius and ulna are separated from the humerus, the biceps and brachialis pull the bones into their proper positions.

Retain the parts *insitu* by the angular splints; support the arm with a pad; and keep the limb quiescent. Nélaton declares that he has seen but one case of this variety of dislocation.

Dislocation of the Ulna Backwards.—This is the only displacement to which the ulna alone is subject; and this seldom presents itself without more or less dislocation of the head of the radius. It may be distinguished by the great deformity of the member,—the olecranon being thrown backwards, and the fore arm and hand very much twisted inwards.

The radius remains in its normal position, and its movements under the hand can be easily recognized when the limb is rotated. The coronoid is frequently fractured in this accident, and crepitates when moved. If this be the case, the dislocation can be *reduced* and *produced* at pleasure.

It is also impossible to extend the arm or to bend it at right angles, in uncomplicated cases of this injury.

Treatment.—Reduction is effected precisely as in the last accident described. The Radius, acts as a lever under these circumstances, and aids the muscles in bringing the bone into position.

Retention may be accomplished by the use of the appliances, &c., described above.

Dislocation of the Radius forwards.—This is a most unusual accident. It may occur however, the result of a fall on the palm of the hand, by which the lower end of the bone is pushed *backwards*, and its upper extremity carried *forwards*, rupturing the annular ligament, and throwing its head against the external condyle. It may be *distinguished* by the following signs, viz: the *forearm* is slightly bent, and can neither be extended nor brought at a right angle with the arm; the hand is fixed midway between *pronation* and *supination*, though neither motion can be perfected; on rotation, the bone can be distinctly felt, and the pain is very great; the whole of the upper side of the arm is carried somewhat *upwards*, producing great deformity; and the constant disposition of the head of the radius to slip out of place because of the rupture of the annular ligament.

Treatment.—Reduce by applying extension after having firmly fixed the upper arm, and then bending the arm and pushing the head into its place.

Retain by applying a pad immediately over the head of the radius, binding it firmly by means of adhesive strips, and keep the forearm well flexed.

Dislocation of radius backwards.—This may be known by the head of the bone being felt subcutaneously *behind* the *external condyle* and by the movements of the elbow being limited and extremely painful.

Treatment.—Reduce by bending the forearm, and making traction.

Retained by keeping the arm flexed. Sir, A. Cooper declared that he had never seen a case of this particular dislocation, in the living body, and **but once upon the dead subject.**

Dislocation of the Radius outwards.—This accident occurs more frequently than the last, according to the testimony of every Surgeon of practical experience. The head of the bone is then on the *outer* side of the external condyle, where it may be felt under the skin, rolling as the hand is moved. The natural motions of the joint are materially interfered with, and pain follows every movement.

Treatment.—Reduction is accomplished by making traction at the wrist, and bending the limb at the elbow.

Retention is effected as in the other cases of dislocation already described.

In *compound* dislocations of the elbow joint, the arm must be flexed and placed in the most comfortable and convenient position, the angular splints applied, when practicable, and the antiphlogistic treatment resorted to. After a few weeks have expired, and the external wound is in good condition an effort may be made to reduce the dislocation.

Dislocations of the Wrist Joint.—Fractures of the lower end of the radius are frequently mistaken for dislocations of the wrist joint, so frequently in fact that some Surgeons have denied the existence of such dislocations under all circumstances. The carpus as a whole may be dislocated either *backwards* or *forwards*. The existence of a *smooth convex swelling* corresponding with the first row of carpal bones either upon the *upper* or *under* surface of the wrist together with some shortening of the forearm, and an unusual prominence of the styloid processes of the radius and ulna, are the guides by which these dislocations may be recognized.

Reduction is readily accomplished by the employment of *extension* and *counter-extension*,—a circumstance which will facilitate the diagnosis between this accident and impacted fracture of the radius.

Retention is effected by means of anterior and posterior splints.

The radius alone is sometimes thrown forwards upon the carpus.

Symptoms.—The outer side of the hand is displaced backwards and the inner forwards, while the extremity of the bone forms a protuberance upon the fore-part of the wrist. *Reduction* and *retention* are effected as when both bones are displaced.

The ulna is sometimes separated from the radius by the rupture of the sacriform ligament, and usually projects *backwards*.

Symptoms.—This accident may be known by an elevation immediately above the level of the *os-cuneiform*, which is easily reduced by pressure to its former situation.

Treatment.—Press the bone back to its proper place, with the finger.

Apply a compress of leather to the extremity of the ulna; place splints along the *forearm*; and use a roller to keep them in position.

Dislocation of the bones of the Carpus.—This accident is of rare occurrence, and is usually the result of falls upon the hand.

The *os-magnum* is the bone most frequently displaced.

Symptoms.—A round hard tumour on the back of the wrist, opposite the metacarpal bone of the little finger, presenting itself immediately subsequent to a fall upon the hand.

Reduction.—Extend the hand and apply pressure upon the tumour.

Retention.—Apply compresses, and enjoin absolute rest.

Instances are on record of the dislocation of the *pisiform* and *semilunar* bones, but these are very unusual accidents.

Sometimes *ganglia* are mistaken for dislocations of these bones, but these are easily removed by striking them sharply with the flat surface of a book, when the supposed dislocation immediately disappears.

A compound dislocation of the carpal bones frequently happens, and is generally produced by the bursting of guns, by the hand being caught in machinery, or by the passing of heavy bodies over it. In such cases one or two of the carpal bones may be dissected away, without destroying the hand or seriously interfering with its motions.

If great injury be done, amputation becomes absolutely necessary.

Dislocation of the Metacarpal bone of the Thumb.—This is the only metacarpal bone that admits of dislocation, and this accident seldom occurs. These luxations have been observed in two directions: *backwards* and *forwards*, and can readily be recognized and reduced—extension being made from the **thumb** by means of a piece of tape applied around **the first phalanx.**

Dislocation of the First Phalanx of the Thumb.—
The bone is usually dislocated backwards but may be thrown forwards also.

Symptoms.—The proximal extremity of the phalanx slides back upon the distal extremity of the metacarpal bone, in the backward dislocation, and stands off from it at nearly a right angle, while the metacarpal bone projects strongly in the palm of the hand. In very rare cases the phalanges are extended upon the metacarpal bone in a straight line.

In the forward dislocation, the first phalanx is in front of the metacarpal bone, and in the same plane: while the last phalanx is inclined slightly back.

Treatment.—If the dislocation be backward, ben the dislocated phalanx forcibly backwards until it stands upon its articulation, hold it in that position, and at the same time press against the distal extremity of the metacarpal bone. Make firm pressure against the base of the dislocated phalanx, and slide it into its place.

If this fail, bend the thumb towards the palm of the hand, in order to relax the flexor muscles as much as possible, and then make extension by means of the clove-hitch. The apparatus of Levis may be also used in this connexion. If the dislocation cannot be reduced by these means, divide one of the short flexors of the thumb, and the reduction can be readily effected. The author is convinced that this is more properly speaking, in many instances, a dislocation of the distal end of **the metacarpal bone, and that reduction can be**
14

most readily ensured by fixing the thumb firmly, and manipulating from the direction of the arm.

When the dislocation is *forward*, reduction may be effected by seizing the thumb in the palm of the hand, and, with the fingers resting upon the back of the patient's hand, forcing the phalanges into flexion by firm and steady pressure.

Dislocations of the *phalanges* of the *fingers* may be reduced on the same principles.

DISLOCATIONS OF THE LOWER EXTREMITIES.—*Dislocations of the Thigh.*—There are four principal dislocations of the femur which should be thoroughly studied and understood by the Surgeon, viz: upwards and backwards upon the dorsum ilii; upwards and backwards into the ischiatic notch; downwards and forwards into the thyroid foramen; and upwards and forwards on the pubes.

1. Dislocation upwards and backwards on the dorsum ilii.

Causes.—Falls from a height when the force of the concussion is received upon the outside of the knee; falls upon the foot or knee when the limb is abducted; a heavy weight striking the pelvis from above, the body being bent forward; or any thing which forces the thigh into extreme abduction or abduction united with rotation inwards.

Symptoms.—The limb is shortened; the thigh is rotated inwards and somewhat flexed; the great toe rests upon the instep of the foot of the sound limb; the knee touches the opposite thigh near the upper margin of the patella; the body of the patient is slightly bent forwards; the roundness of the hip is lost; the trochanter major is depressed; and the head of the bone can be felt in its new position.

Treatment.—The dislocation may be reduced by manipulation or by mechanical force, (extension and counter extension).

Manipulation.—Hippocrates first described this method of reduction, though it has been variously modified, illustrated and improved by Wiseman, Turner, Anderson, Physic, Smith, Colombat, Reid and others.

Directions.—Place the patient in the horizontal posture on a narrow table covered with blankets, and on his sound side. Secure the body firmly by folding a sheet several times lengthwise, then apply the middle of the band thus made, to the inner and upper part of the sound thigh, carry its extremities under the table, pass them obliquely up, cross them again firmly over the trunk above the injured hip, and secure the ends under the table.

Administer chloroform freely; stand at the patient's back; grasp the knee of the dislocated limb with the right hand and the ankle with the left—if the left femur be dislocated reverse the hands; flex the leg upon the thigh; rotate the thigh outwards; then *slightly* abduct the thigh by pressing the knee outwards; and lastly thrust the knee upwards towards the face, so as to flex the thigh freely, and at the same moment increase the abduction of the limb. This is the plan of Nathan Smith, as described by his son, the distinguished **Professor of Surgery in the University of Maryland.**

Mechanical Means.—Reduction by extension dates back to Hippocrates, but Ambrose Paré was the first to recommend the use of pullies. The plan to be pursued in this connexion is as follows: place the patient upon a bed of suitable height, on his back and slightly turned on the sound side; drive a staple into the wall of the room upon one side and another into the wall upon the opposite side, both corresponding with the line of the shaft of the femur, but the one in front being higher and the one behind being lower than the bed; lay two pieces of strong cloth, four inches wide and four feet long, on either side of the limb, the centre of each being just above the two condyles; over the centre of these two strips apply a strong roller tightly, previously wetted in water; bring down the upper ends of the side strips and fasten them to the lower, so as to form two loops, upon which one of the hooks of the compound pulley is to be made fast, while the other hook is secured to the front staple in the wall; fold a sheet diagonally, and adjust it so that its centre applies to the peritoneum while its ends are tied to the lower staple; pass underneath the upper part of the dislocated limb, a strong, broad bandage of sufficient length to tie over the neck of the Surgeon when standing about half bent; place assistants on either side of the patient to keep him in position; everything thus prepared, *administer chloroform*, make extension by means of the pulley, and counter extension by means of the sheet, in the line of the axis of the dislocated limb; place the hand carefully upon the trochanter major, and watch carefully its descent; and then when the head of the bone has nearly or quite reached its socket, if it does immediately get into position, lift up the thigh by means of the hand, which has been passed under it, and the luxation will generally be reduced.

If, after all, the bone does not enter the socket, the flexion of the limb may be increased or diminished, the tension suddenly released, and "manipulation" attempted·

The extending force may be applied also by means of a leather belt, strips of adhesive plaster, &c; while a small rope doubled upon itself, with a stick passed through it, may be substituted for the pulley. Bloxham, "dislocating tourniquet," and Jarvis' adjuster may also be employed in this connexion.

2. Dislocation upwards and backwards into the great ischiatic notch.

Causes.—Falls upon the foot or knee, when the limb is very much in advance of the body; heavy blows upon the back and pelvis when the thigh is nearly at right angle with the body, &c.

Symptoms.—The limb is shortened, but not so much as in the last named dislocation ; the thigh is flexed, adducted and rotated inwards ; the toe of the dislocated limb touches the ball of the great toe on the other side; the knee is not carried so far over the other as in the former luxation ; the trochanter major is approximated towards the anterior superior spinous process of the ilium; and the lumbar part of the spine is so arched that it cannot be straightened so long as the thigh is straight or on a line with the patient's trunk.

Treatment.—Manipulation may be employed, precisely as described above, though, the extent of the circuit to be described by the head of the bone is inconsiderable, while there is great danger of its being thrown into the foramen thyroideum.

Extension.—Arrange every thing as before described, taking care to have the "front staple" at a greater height from the floor; administer chloroform; make extension at an angle of 45°; and when sufficient force has been applied lift the thigh upwards by means of the band passed under the thigh and carried over the operator's shoulder. Bransly Cooper says that the limb should be flexed quite to a right angle while extension is being made.

Be careful that the "counter extending" band does not slide off the pelvis toward the upper part of the thigh.

3. Dislocations downwards and forwards into the foramen thyroideum.

Causes.—Falls upon the foot or knee while the limb is abducted, and the falling of a heavy weight upon the back the body being bent and the thighs spread asunder.

Symptoms.—The thigh is lengthened one or two inches, abducted, flexed, and advanced; the body is bent forwards or slightly flexed upon the thigh; the toes point directly forwards as a general thing; the hip is flattened; the trochanter is less prominent; and the head of the bone may be often felt in its new position.

Treatment.—Manipulation. This dislocation may be readily reduced by manipulation if conducted in the following manner; abduct the limb; carry it up towards the body until the progress of the knee is arrested; then carry the limb inward ; and finally bring it down adducted. When the knee is opposite the pubes, rotate the femur quickly inwards, and give it a slight rocking motion. Extension: Sir A. Cooper advises that exten be made in the following manner; place the ient on his back with thighs separated; make the leys fast to a band drawn through the perineum of the affected side, in a direction upwards and outwards; pass a counter band around the pelvis through the band attached to the pulleys, and attach it to a staple driven in the wall; administer Chloroform; make traction with the pulleys until the head of the bone is felt moving from its position; then seize the ankle and adduct the limb forcibly. Place the patient in bed and rotate the limb inwards, keeping the knees together.

4. Dislocation upwards and forwards upon the pubes.

Causes.—Falls upon the foot, when the leg is thrown backwards; putting one foot into a hole while walking and falling backwards; and falls or blows upon the back of the pelvis.

Symptoms—The thigh is shortened, flexed slightly and rotated outwards; the trochanter cannot be distinguished; the head of the bone can be felt on the pubes or outside of the femoral artery.

Treatment.—Manipulation. Numerous instances of the reduction of this dislocation by manipulation, are on record, though the methods pursued were different. The best plan, is as follows, ab-

...t the limb and forcibly rotate it outwards; flex
...pon the body; then adduct it, and bring it
...wn upon the table. Care should be taken not
...continue the rotation outwards after the head
...the femur has risen above the pubes, but on the
...ntrary to rotate it gently inwards so as to enable
...e head to slide under the psoas and iliacus in-
...rnus muscles towards its socket. Extension.
...ay the patient on his back upon the table; make
...e extending band fast above the knee and attach
...o a staple driven in the floor; pass the counter
...ending band under the perineum and attach it
...staple above the level of the table; administer
...roform; make steady and persistent extension;
...when the head of the femur has begun to move,
...he upper part of the thigh, as before described,
...to carry the head of the bone into its socket.
...here are three cardinal principles which should
...membered in this connexion, viz:

In reducing by manipulation, carry the limb
in those directions in which it is found to
...easily.

In reducing by extension apply the force in
...tion of the axis of the dislocated limb.

...r reduction has been effected, particular-
...uch force has been used, keep the patient
...ely in bed, with his knees brought together
...t... all danger of inflammation and recurrence of
...e accident, have passed.

Various other anomalous dislocations may occur
...this connexion, for an account of which the
...is referred to the standard works on the
subject.

Dislocations of the Patella.—This bone may be dislocated either outwards, inwards, upwards or upon its own axis.

Causes.—Muscular action of a sudden and spasmodic character; blows; falls, &c.

Symptoms.—The altered position of the bone; the prominence of either condyle; the immovable condition of the limb; great pain; and slightly bent condition of the knee.

Treatment.—The treatment consists in relaxing the quadriceps extensor muscle, in extending the leg, in carrying the body forward, and then pressing the bone into position.

Dislocation of the head of the Tibia.—The head of the Tibia may be dislocated backwards, forwards, inwards, outwards, and backwards and outwards, though the accident is of rare occurrence.

Dislocation of the head of the Tibia backwards. *Causes.* Violent blows upon the lower end of the femur or upper end of the tibia; and by the twisting of the tibia when the foot is made fast in a hole and the body swings around upon the knee.

Symptoms.—The head of tibia may be felt in popliteal space pain in consequence of pressure upon the popliteal nerve; a depression immediately below the patilla; the condyles of the femur project; and the limb usually somewhat flexed.

Treatment.—Manipulation may succeed if the injury be very recent or the shock great. The limb should be carried in those positions in which it moves most easily; but if this fails then forced flexion should be resorted to, rocking the limb from one side to another, and making strong pressure upon the projecting bones of the joint. Extension may be practised by making a strong assistant seize the limb above the ankle, and pull forcibly in the direction of the axis of the limb. The pulleys may also be employed. Counter extension may be made from the perineum, or from the lower and under part of the thigh. Dislocation forwards. The *causes* by which this accident is produced, are similar to those mentioned above.

Symptoms.—The patella, fibula and tibia are prominent in front, while the condyles of the femur may be felt behind; the limb is shortened; and the circulation is interrupted by pressure upon the artery.

Treatment.—Manipulation may possibly succeed if attempted immediately. Extension and counter extension should be made as described above.

Dislocation outwards.—*Causes*. A violent wrench of the knee joint, may rupture the ligaments, and cause this accident.

Symptoms.—The inner condyle of the femur projects, while the head of the tibia and fibula can be distinctly felt on the outer side of the joint.

Treatment.—The treatment does not differ from that of the other dislocations just described.

In the dislocations inward and outward and backward there is nothing peculiar, and the accident, should be treated on general principles.

Dislocations of the Lower end of the Tibia.—The tibia may be dislocated at its lower end in four directions, namely: Inwards, outwards, forwards and backwards. Most of these accidents complicate themselves with fractures, of the two bones of the leg, one or both.

Dislocation inward, *Causes*. Falls from a height upon the bottom of the foot, which at the same time is turned outwards; blows, and violents twists of the foot outwards.

Symptoms.—Foot is abducted; the internal malleolus projects strongly; there is a corresponding depression upon the outer side of the ankle; the pain is great; motion is lost, though the surgeon can move the foot; and, fracture of fibula when the dislocation is complete.

Treatment.—Seize upon the foot, and forcibly adducting it, taking pains to flex the leg so as to relax the gastrocnemius muscle, and to give the part a gentle rocking motion. If this fails, bend the leg up a right angle to the thigh; pass a counter extending band around the thigh; attach the pulleys to the foot by means of a bandage carried around it; and then make forcible extension.

Dislocation outwards.—*Causes.*—The causes are similar to those which produce the last named accident, only the position of the foot is reversed.

Symptoms.—The foot is adducted; the external malleolus projects; there is a depression upon the inner side of the foot, &c.

Treatment.—The outward dislocation may be reduced precisely in the same manner as the dislocation inwards.

Dislocation forwards.—*Causes.*—Violent extension of the foot upon the leg; falls upon an inclined plane; blows upon the tibia, &c.

Symptoms.—The length of foot in front of tibia is

diminished, while the projection of the heel is increased; the toes are turned downwards; the heel is drawn upwards; the end of the tibia can be felt; and the tendo-Achillis is curved forwards and tense.

Treatment.—Flex the leg upon the thigh, make extension from the foot; and at the same time, press in front of the tibia and against the heel. When the bone begins to slide into its place, the foot should be forcibly flexed upon the leg.

Dislocation backwards. This is so rare an accident that Malgaigne has only succeeded in collecting five examples. It is produced by causes exactly the reverse of those which operate in the production of the last, while the signs which distinguish it are directly opposite to those last described. Reduction should be attempted by a method similar to that recommended for other dislocations of the ankle joint.

The Fibula may also be dislocated both at its upper and lower end, but these accidents are of such rare occurrence, and so readily distinguished as to preclude the necessity for a more detailed account of them.

Dislocation of the Astragalus.—*Causes.*—The same as those which produce dislocation of the Tibia.

Symptoms.—Prominences according as the bone is displaced inwards, outwards, backwards or forwards; lateral deviation of the foot; shortening of the leg, &c.

Treament.—Reduce if possibly by means of extension pressure, &c., but if unsuccessful, resect or amputate. Keep down the inflammation, which is always intense.

The Astragalus may also be separated from the Scaphoid bone, and should be treated on the same principles.

Dislocation of the Calcaneum.—Causes.—Falls upon the heel and direct blows.

Symptoms.—Prominences and depressions according as the dislocation is outward, upwards and inwards.

Treatment.—Bend the thigh and knee on the body; flex the leg; seize the metatarsus and the heel; draw the foot directly from the leg; and press the knee against the outside of the joint.

The Scaphoid, the Cuneiform bones, the os-cuboides and metatarsal bones are all subject to dislocations, which can be recognized without much difficulty and which should be treated on the same general principles as the bones of the foot already referred to above.

CHAPTER VIII.

FRACTURES.

FRACTURES IN GENERAL.—The term *fracture* is derived from a Greek word which signifies "to break," and is employed to convey the idea of a division, by violence, of bone or cartilage.

Classification.—The following is the most simple and convenient classification of fractures:

All fractures are:

INCOMPLETE	OR	COMPLETE.
Embracing.		*Embracing.*
Fissures,		Transverse fractures,
Depression,		Serrated "
Curvature,		Oblique "
Flexion,		Impacted "
Splintering,		Stellated.
Perforations.		

EITHER OF WHICH MAY BE:
Simple,
Compound,
Comminuted,
Complicated.

Incomplete Fractures.—These involve the division of only a portion of the thickness of the bone, and embrace.

1. *Fissures.*—The experience of all surgeons confirms the fact that both flat and long bones may be *cracked*, in any direction as the result of violence, without a solution of their entire continuity. The symptoms which mark this accident are those of contusion of the bone, and depend upon the development of periostitis, or of suppuration in the medullary canal or internal structure of the bone.

2. *Depression.*—This term is employed to designate the circumscribed fracture of a part of the thickness of a flat bone with more or less flexion of the portion which remains intact. Depression has been observed in the bones of the cranium, ribs, scapula, neck of the femur, and of the diaphyses generally. This accident can readily be determined, in a majority of cases, by thrusting the finger into the depressions.

3. *Flexion.*—The long bones may all be bent in the direction of their diameter, as the result of a similar lesion. Under these circumstances there is not simple curvature, but positive fracture of a portion of the thickness of the bone, save in the case of very young subjects.

This accident occurs most commonly in the bones of the fore arm; then in the thigh; and lastly in the leg. The young—those between the ages of five and thirteen—are more subject to it than persons of mature years.

The bone is generally more or less curved, with a salient angle on the side of the fracture; while the curvature can be diminished but rarely overcome by pressure.

4. *Splintering.*—There may be a complete sepa-

ration of a mere splinter while the bone itself remains nearly solid. Fractures of this description are usually produced by blows of a sabre, or by falls grazing the bone, and may occur in any part of the body, though the skull is most frequently the locality of the accident· The splinter can usually be felt and the diagnosis is not difficult.

5. Perforations.—The bone may be perforated through and through or in one portion of its thickness by foreign bodies, particularly by balls, without the complication of splinters or comminution. In the one instance the perforation is said to be *complete* and in the other, *incomplete.* These lesions have been observed in all the bones of the body, and are of constant occurrence, though true perforations occur most frequently in the bones of the skull, and the head of the femur and tibia. These accidents are generally serious. The surrounding soft parts swell and inflame ; the bone also takes on inflammatory action : the limb becomes œdematous ; a fœtid reddish pus flows from the wound; while a probe introduced into it shows that the bone is soft and easily broken down. The splinters are detached and float out with the purulent matter ; and either the work of repair is commenced, or caries is developed, fistulæ are produced, a tedious suppuration ensues, and amputation or resection becomes the only available remedy. The great indication is to extract the foreign body, in the premises, if the perforation be an incomplete one. The wound should then be detached on general principles.

Complete Fractures. When the bone is divided

COMPLETE FRACTURES. 341

to the extent of its whole thickness, the fracture is said to be complete. Fractures of this kind are:

1. Transverse. When the line of fracture forms a right angle with the long diameter of the bone, or deviates from the perpendicular so slightly as permits the ends of the bone to rest upon each other, or when replaced not to become spontaneously displaced, the fracture is transverse.

2. Serrated Fractures.—When the opposite surfaces denticulate, the elevations upon one fragment being reflected by corresponding depressions in the other, the fracture is serrated. A majority of fracture from simple blows are of this character; but they occur principally in the clavicle, humerus, radius, ulna, femur, and tibia.

3. Oblique Fractures.—When the line of fracture forms an angle with the shaft of the bone not far from 45°, the fracture is oblique. When the obliquity is less than forty-five degrees, the fracture becomes transverse, when greater, it is styled a fracture *en bec de flûte*, and when it approaches parallism to the axis of the bone, it is called a longitudinal fracture. These fractures are generally produced by indirect violence, and usually have something peculiar in their aspect, according to the cause producing them.

4. Impacted fractures.—When the ends of the bone are driven into each other, the lamellated structure of one fragment penetrating the cancellous structure of the other, the fracture is said to be impacted.

5. Stellated Fractures.—When some cutting in-

strument or a ball is driven through the bone, particularly if it be a flat one, innumerable spiculæ will in many instances be found projecting from the margins of the perforation. The projection or radiation of these fragments from a central point gives the fracture a stellated or star like appearance; and hence, the name of the fracture. This accident occurs frequently in connexion with the bones of the cranium, most seriously complicating those accidents.

In addition to these distinctive characteristics, there are some features which may connect themselves with either incomplete or complete fractures. Thus both varieties of fracture may be either simple, compound, comminuted or complicated.

1. Simple Fracture.—By this term is usually meant the fracture of a bone simply at *one point*, without reference to the question of complications. A more correct and convenient arrangement would extend its meaning thus: "a fracture simply at one point without injury to the soft parts."

2, Compound Fracture.—When there is an external wound communicating with a fracture of the bone, whether complete or incomplete, the injury is recognized as a compound Fracture.

3. Comminuted Fracture.—When the bone is broken at more than one point, and there are more than two fragments, the fracture is "multiple" or comminuted.

4. Complicated Fracture.—A fracture is said to be complicated, when in addition to the division of the bone, there is injury either of some impor-

tant vessel or nerve, great contusion or laceration of the soft parts, fracture of neighbouring bones, dislocation, or constitutional injury.

Causes of Fractures.—The causes of fractures are predisposing and exciting.

Predisposing Causes.—In childhood the bones are soft and easily bent, and in old age they are harder and more brittle. Females are less liable to fracture than males except in old age. More fractures occur in winter than in summer. Mollites Ossium, Fragilitas Ossium, Rickets, Cancer, Syphilis, Scrofula, Gout, Scurvy, Mercurialization, &c., all predispose to the occurrence of fractures.

Exciting causes.—The exciting causes of fracture are mechanical violence, and muscular action. Mechanical violence, may act either directly or by counter stroke. Muscular action most frequently produces fractures of the patella, calcaneum, humerus, femur, tibia, and olecranon process of the ulna, and usually implies some predisposition to the accident.

General Symptoms of Fracture.—The most common and important signs are crepitus; mobility; inability of the parts to remain in position when reduced; pain at the seat of fracture; swelling; ecchymosis; deformity; and inability to move the limb. The examination of a suspected fracture should be made as early and as quickly as possible, Chloroform being employed if there is the least difficulty in regard to the diagnosis.

Treatment of Fractures.—The treatment of fractures divides itself, naturally into two processes, viz: reduction and retention. Before discussing them however, it will not be amiss to consider the manner in which a man who has sustained a serious fracture should be cared for in advance of regular Surgical treatment. If the upper members are broken, the patient, as a general thing, can take care of himself, but when the inferior extremities are involved, he should be most tenderly and intelligently cared for. An army should not only be well supplied with litters, but a permanent detail should also be made from each regiment of some of its bravest and strongest men, of the same height, to manage them. This "litter-corps" should be under the immediate direction of the assistant Surgeon,—to be instructed by him before the fight, and directed by him during the engagement, so that immediate and proper attention may be given the wounded in the first moments of their suffering and danger. The advantages of this arrangement are two fold; it prevents those who are uninjured from leaving the ranks under the excuse of taking care of the wounded, and it secures prompt assistance for those who have been mutilated by the bullets of the foe. The importance of prompt and proper assistance is particularly apparent in connexion with fractures of the lower extremities,—in as much as it precludes farther displacement of the fragments, additional injury to the surrounding soft parts, a more serious shock to the system, and unnecessary pain to the patient himself. This sad picture is too often presented on the field of battle, —the wounded are put astride guns, raised by their garments, or rolled up in a blanket and dragged to a place of safety. When a proper litter cannot be had, one may be made from the blanket of a soldier, according to the directions of Chisolm. Thus, double the blanket upon itself; make a slit through the end corners sufficiently large to admit the barrel of a musket; then pass one musket through the fold of the blanket and another through the slit in the ends. This is very defective, and if the door or blind of a house can be procured, it may be substituted with advantage.

The easiest way to raise a patient and place him on a litter is this: a strong man standing on the sound side, puts one arm round the patients chest, and the other hand under the pelvis, while the patients arm is placed around his neck; another assistant should support the pelvis, and a second the sound limb; two others sustain the broken limb by its two extremities, taking care to keep it in the straight position; and at a given signal the patient is raised and the litter slipped under him. In the same way he may be transfered from the litter to his bed or from one bed to another. The patient having been carried to the Infirmary or Hospital and divested of his clothes, by the most careful manipulation, should then be subjected to treatment.

Reduction.—By this term is meant the bringing of the bones in proper position. It is accomplished by means of extension and counter extension—forces which are employed to overcome the muscular contraction by which the fragments are kept in an unnatural position, or by relaxing the muscles, and at the same time so manipulating as to bring the bones properly in contact.

Retention.—The bones must not only be returned to their normal relations but kept there by some mechanical contrivance. This object is accomplished by means of bandages and splints, so adjusted as either to overcome the muscular contraction which operates as the separating force, or by relaxing the muscles exerting that force upon the fragments, to secure their retention in the proper position. This whole matter will be more fully explained in connexion with particular fractures.

The treatment of compound and complicated fractures is a matter of particular moment to the military Surgeon, especially since the introduction of conical balls. The following considerations should serve as a guide, in this regard.

1. The upper limbs manifest a much greater vitality and resistance to injuries than the lower, so that extensive injury to their bones does not necessarily demand amputation.

2. The effect of a conical ball upon a bone is nearly always frightful, either comminuting it to a great extent, or splitting it longitudinally, even when the soft parts are but slightly injured.

Whatever then may be the indication given by the external wound, it is the duty of the Surgeon, to explore it thoroughly and without delay, for the purpose of ascertaining to what extent the bone is injured, and of deciding upon the best mode of treatment.

3. The question to be decided by the Surgeon is not, whether it is *possible* to save the particular case before him, but whether, with the appliances

and facilities at his command it is *probable* that conservative Surgery would ensure a favorable result. In civil practice, when the patient can be made comfortable and is surrounded by proper hygienic conditions, an attempt should generally be made to save the limb; while in military Surgery, where discomfort is a necessity and a crowded Hospital teeming with the putrescent emanations of filthy wounds, the only available receptacle for the mutilated victim, then, an immediate amputation should be resorted to.

4. The fracture of a bone by a conical ball is always attended with frightful sequelæ. Violent inflammation is speedily developed accompanied by great pain, swelling and nervous shock, and followed by extensive and protracted suppuration. The ability of the patient to stand so serious a commotion and so great a drain should likewise be taken into the account.

5. In the event of an attempt being made to save the limb, the Surgeon should proceed to remove all loose spiculæ; to smooth off the sharp and irregular ends of the bones; to place the limb in the most comfortable and convenient position; to apply cold water or iced bladders to the wound; to promote the escape of pus; to relieve pain, as far as practicable; to support the strength of the patient; and when the inflammatory action has subsided, to apply such apparatus as is best calculated to prevent deformity.

When union does not take place within a reasonable time it may be facilitated in various ways. Thus blisters, caustics, electricity, mercury, the seton, loops of wire, acupuncture needles, abraiding or removing the ends of the bones, and subcutaneous puncture have all been recommended in this connexion. Brainard employs a strong metallic perforator, so hardened as to penetrate the hardest bone or ivory, which he employs in the following manner: In case of oblique fracture or one with overlapping, the skin is perforated with the instrument at such a point as to enable it to be carried through the ends of the fragments, to wound their surfaces, and to transfix whatever tissue may be placed between them. After having transfixed them in one direction it is withdrawn from the bone, but not from the skin, its direction changed and another perforation made, and this operation is repeated as often as may be desired. This is perhaps the best of all the plans devised to fulfill the indications in such cases. In conjunction with it, the condition of the patient's general health should be improved; all local impediments removed; the action of subjacent tissues promoted; and the patient allowed to walk about.

PARTICULAR FRACTURES.—*Fracture of the Cranial Bones.*—All the bones of the head may be fractured either by direct violence or *contre-coup*. These fractures are incomplete or complete, either variety of which may be simple, compound or complicated.

Incomplete Fracture.—The forms in which this lesion presents itself, are, (1.) Cracks or fissures of either the outer or inner table, there being no depression or separation of the bones; and (2) depression of the bones without a solution of their continuity.

1. These cracks or fissures may be associated either with mere contusion of the scalp, or with a wound of it. As a general thing there are no particular signs by which the accident can be distinguished, though there may be associated with it, primarily, concussion, and secondarily compression or inflammation of the brain.

2. Depression may occur in young children without a positive division of the structure of the bones, having the same associations and complications as were referred to in connexion with cracks or fissures.

Complete Fractures.—This variety of fracture may present itself under the following forms, (1) fracture without depression; (2) fracture with depression; (3) fracture with comminution of the bone; (4) fracture with a removal of a portion of the bone, as from sword cuts; (5) fractures with depression of both tables, and great splintering of the internal one, as from bayonet thrusts or even from the effects of musket balls.

1. Simple fracture without depression. Both tablets may be divided, but not displaced, in connexion either with mere contusions of the scalp or with wounds of it, as the result of direct violence or of a contre-coup. This accident can generally be determined by running the finger nail or the end of a probe over the exposed surface of the wound or by seeing a fissure into which the blood sinks. The most serious and fatal form of simple fracture is that which extends through the base of

15

the cranium, and in a majority of cases is the result of counter stroke. The strongest presumptive signs of the existence of this injury are the escape of blood or of a serous fluid from the ears and nose. If the bleeding be persistent and symptoms of serious injury to the brain speedily follow a severe blow or fall upon the head, the occurrence of such a lesion may be suspected.

Even simple fracture may be complicated with concussion, compression or inflammation of the brain.

2. Fracture with depression. Depression may associate itself either with a simple, a compound or comminuted fracture. A portion of the bone is found depressed or driven beyond its level to a greater or less extent according to the nature of the accident. The symptoms or signs which distinguish this accident are of two kinds; those which are dependent upon the injury of the bone and those which result from the concomitant pressure and laceration of the brain.

As a general thing the depression can be felt, particularly if there be an external wound; but the surest proof of the occurrence of the injury consists in the immediate manifestation of symptoms of compression of the brain, and the subsequent development of the characteristic phenomena of cerebral inflammation. In order that the pathological difference between concussion and compression of the brain, may be thoroughly comprehended, the essential phenomena of each are here tabulated.

CONCUSSION.	COMPRESSION.
1. The symptoms are usually immediate.	1. The symptoms are usually delayed for a few moments.
2. The patient can be made to answer questions incoherently.	2. The patient cannot be roused, and is speechless.
3. The patient can still hear, see, taste and feel.	3. Special sensation is destroyed.
4. Respiration is feeble and noiseless.	4. Respiration slow, laborious, stertorous, and blowing.
5. The pulse is weak, tremulous, intermittent and frequent.	5. The pulse is slow and full.
6. There is nausia, and vomiting.	6. The stomach is quiet and insensible to emetics.
7. Bowels are relaxed.	7. The bowels are constipated.
8. Water sometimes flows from the bladder, but is usually voided regularly.	8. The bladder is paralysed, and the use of the catheter is necessary.
9. There is no paralysis of the muscles.	9. There is always paralysis, and on the opposite side from the wound.
10. The pupils are irregularly contracted or dilated.	10. The pupils are widely dilated.
11. The brain is shaken.	11. The brain is compressed.
12. The surface is pale and cold.	12. The surface is not pale and cold, but rather the reverse.
13. The pulse grows stronger as normal condition returns, reaction and fever ensuing.	13. The pulse grows weaker as health returns—collapse frequently ensues.
14. The symptoms indicate a condition of syncope.	14. The symptoms indicate a state of cerebral apoplexy.

Concussion may terminate in compression in consequence of the pouring out of blood from the small vessels which have been divided by the oscillation of the brain, after reaction has taken place; while either of these conditions may even tuate in the development of inflammatory action, with its characteristic phenomena and terminations. Compression may be caused by either or all of the following causes: depressed bone; extravasated blood; and purulent deposit—the first producing its effects immediately, the other after the lapse of some little time, and the last, at a more remote period in the history of the case.

3. Fracture with comminution of the bone.—The skull is frequently the seat of multiple fractures. Either from the abnormal condition of the bones or the peculiarity of the injury, it often happens that the bones are broken into a number of fragments. Balls usually pass through the bones without splintering them, but it sometimes occurs that the injurying force disseminates itself, producing extensive comminution of both tablets.

Under these circumstances the danger is from inflammation of the brain and its membranes; and from fungus of the brain. There may be depression with its usual symptoms, but, as there is extensive solution of the bony continuity, the fragments are not held down upon the surface of the brain by any considerable force, and not unfrequently rise again to their original level, relieving the cerebral substance of the disasterous consequences of their presence. Many of the detached fragments become necrosed, dying slowly and endangering the delicate structures beneath them until the work of elimination has been perfected. The substance of the brain may be injured contemporaneously with the fracture of the bone, causing a speedy extravasation of blood, and the development of cerebral inflammation.

4. Removal of a portion of the bone. Portions of the skull are sometimes carried away by sword

cuts, which if promptly reapplied will adhere without unfavorable consequences. There is always danger of hernia and inflammation of the brain.

5. Fractures with splintering of the internal table. These accidents result from bayonet, dirk and ball wounds, and are of the greatest interest to the Surgeon. They are always complicated with injury to the cerebral surface, and frequently, by the actual presence of a foreign body in the wound. While connected with the General Hospital at Charlottesville, Va., it fell to my lot to make an autopsy of a soldier who had died from the effects of a gunshot wound of the head. A conical ball had entered the left parietal bone about one inch from the sagittal suture, making a smooth, round hole in the external tablet, and imbedding itself in the substance of the brain. The patient was attacked with violent convulsions on the fourteenth day subsequent to the receipt of the injury, and died comatose in a few hours afterwards. Upon removing the upper half of the cranium, a large abscess was found immediately beneath the orifice in the bone, containing the ball and a quantity of puss; and the whole dura mater was injected with b ood; while at the point of the inner tablet through which the missile had passed, was a round hole, from the entire circumference of which there radiated numerous spiculæ which had penetrate the membranes of the brain and acted as foreign and offending bodies to them, as well as to the de-cate structure beneath.

This case is but a type of hundreds of others, and throws much light upon the pathological conditions which such injuries develop. Whatever may be said of the trephine, it is plain that it might have been employed to advantage in this connexion, for the following reasons.

1. It would have ensured the removal of the ball.

2. It would have accomplished the evacuation of the pus.

3. It would have removed the spiculæ which were sources of inflammatory disturbance to the cerebral substance. As a matter of curiosity, an attempt was made to remove the spiculæ which radiated from the circumference of the inner orifice, in order to determine to some extent how far such a procedure could be regarded as "meddlesome Surgery" in actual practice. By means of a delicate pair of forceps, and with a little care, all of them were speedily removed through the external orifice of the wound—the whole thing being accomplished with so much facility as to convince all present of the practicability and propriety of such an operation under *any* circumstances, provided the opening be large enough.

When the fracture has been occasioned by puncture with a sharp instrument, the lesion should be esteemed one of importance and gravity.

The danger is from cerebral inflammation, which ensues within a few days, and generally destroys life. This accident can always be recognized by digital compression aided by a probe, particularly if the previous history of the case can be obtained.

Treatment.—In fissures and simple fractures without local or general complications, keep the patient quiet; give a mild purgative; and apply cold applications to the seat of injury.

Guthrie has wisely remarked that "injuries of the head affecting the brain are difficult of distinction, doubtful in character, treacherous in their course, and for the most part fatal in their results;" while Macleod declares that, of all the accidents met with in the field, these are the most serious, both directly and indirectly—the most confused in their manifestations and the least determined in their treatment." In the truth of these observations all military Surgeons must agree, since this class of injuries still constitute the opprobrium of their art notwithstanding the researches and labors of its ablest masters.

A remarkable disparity presents itself between the injury inflicted and the effects produced by it. Thus, in many instances wounds, apparently of the most trivial character, are followed by the gravest results; while in other cases, extensive comminution of the bones of the cranium, together with considerable destruction of the cerebral substances itself, produces but an inconsiderable disturbance in the economy. As regards the prognosis in this connexion, the rule is to hope for everything, whatever the nature of the injury, but to be confident of nothing, since recovery may follow the gravest accident and death ensue upon the slightest.

Cunningham relates the case of a boy who lived for twenty four days with the breech of a pistol weighing nine drachms lying on the tentorium and resting against the occipital bone. O'Callaghan has recorded the case of an officer who lived seven years with the breech of a fowling piece, weighing three ounces, lodged in the forehead and in contact with the brain. Ellerslie Wallace, gives the case of a girl who rapidly recovered without an unto ward symptom, from a wound inflicted by a circular saw, four inches and a quarter in length, by one sixth of an inch in width, extending across the skull, wounding the brain, and dividing the longitudinal sinus. Hennen states that he has seen five cases in which bullets were lodged in the brain without proving immediately fatal; and also metions an instance in which the bone was depressed in a "funnel shape" to the extent of an inch and a half, without producing an unfavorable symptom. The most remarkable case is that reported by Bigelow, in which, by the premature explosion of a blast, a tamping-iron, three feet four inches in length, one and a quarter inches in diameter, and weighing thirteen and a quarter pounds, traversed the cranium from the angle of the lower jaw on one side to the centre of the frontal bone above, near the sagittal suture. From this extraordinary lesion the patient recovered, with the loss only of the sight of the injured eye.

The effects of an injury inflicted by balls striking the skull will depend upon the following circumstances:

1. Upon the manner in which the ball strikes the skull. When the direction of the projectile is very oblique, and its force considerably exhausted, the injury inflicted may be only a slight contusion of the soft parts or of the bone. When the force is greater, the scalp may be extensively lacerated

and the bone bruised and broken throughout its whole extent, or through one of its tables only, and the cerebral substance beneath considerably injured. Under these circumstances concussion is likely to ensue, terminating, it may be, in "encephalic inflammation and compression from effusion."

Again, a shot which merely grazes the head and "brushes" over the skull, may completely smash the bones of the cranium without injuring the scalp, or by only opening the veins immediately beneath the skull, produce instant death.

2. *Upon the character of the ball.*—Conical balls crush through both tables, with great violence, producing orifices of *equal seize*, comminuting the bone extensively, and carrying the fragments deep into the substance of the brain. Round balls, on the contrary, neither produce so great a destruction of the outer table, nor so extensive and minute a comminution of the bones.

The greater splintering of the *inner* than of the *outer* table, which usually occurs in wounds of the head, is explicable by the fact that the latter is better supported by the parts beneath it, and that the momentum of the ball is necessarily diminished in passing through them. The same principles interpret the difference between the wounds of entrance and exit in the soft parts. In this connexion it may be well to sum up the differences by which the wounds of entrance and exit, in the soft parts, can be distinguished.

THE WOUND OF ENTRANCE IS:	THE WOUND OF EXIT IS:
1. Regular and inverted.	1. Irregular and everted
2. White, depressed, and adherent to the underlying parts.	2. More discolored, but indistinct and not adherent.
3. Characterized by positive loss of substance, and sometimes by the presence of foreign substances, as clothing, &c.	3. Characterized by a flap like tearing, and by no complication of foreign substances, &c.
4. More disposed to bleed than the wound of exit.	4. Less disposed to bleed than the wound of entrance.

These differences are by no means constant and invariable. The speed of the ball, the mode of impingement, the nature of the wounded structure, and the distance at which the gun is fired, exercise a material influence in determining and modifying their character.

The great velocity and peculiar motion of conical balls impress upon wounds a character materially different from that caused by round balls.— When the distance is short, and the parts fleshy, there is less laceration of the soft parts; but "when the range is greater and the part struck bony, the tearing especially at the place of exit is greatly more marked."

They, also, may lodge beneath the outer table without penetrating the cranium, or, after striking against a bony angle or projection, split into two fragments,‡ one entering the skull and the other flying off; and again, in some instances, they have been known to be deflected from their course, af-

---

‡ Macleod, in speaking of this subject, refers to the fact that round balls frequently split, but remarks that he does not believe that "the conical ball with its immense force of propulsion could be so split." In the clothes of a friend of the author, who was killed at Malvern, one half of a conical ball was found, the other half having penetrated the body and produced his death.

ter dividing the scalp, and, without fracturing the bones, to make the entire circuit of the head.

3. Upon the part struck.—Wounds of the side of the head, especially anterior to the ear, are the most dangerous,—thus, a descending scale will give the following order: the fore part, the vertex, and the upper part of the occipital region. Wounds of the base of the brain, especially of the pons and medulla, are necessarily and immediately fatal.

When large vessels are divided, especially the sinuses, death takes place as a matter of necessity.

4. Upon the age and temperament, &c., of the patient.—In the young the same danger is not to be apprehended from injuries of the head as in adults.

It can also be readily understood how the temperament, and the general surroundings of the sufferer exert an influence upon the prognosis, by increasing or diminishing the tendency to control inflammation, and assisting or interfering with the proper treatment of the case.

Shell wounds, though comparatively rare, produce the most fearful injuries and speedily terminate in death.

For all wounds of the scalp, danger from erysipelas is to be apprehended.

STATISTICS.

Macleod reports 630 cases of mere contusion, with 8 deaths; 135 cases of fracture with depression, with 76 deaths; 67 cases of penetration, with 67 deaths; and 19 cases of perforation with deaths. Alcock reports 28 cases of fracture with gunshot wounds, with 22 deaths. Maniere reports 10 penetrating wounds, with 10 deaths. Lente reports 128 cases of fracture of the skull, with 106 deaths. Stromyer reports 41 cases of gunshot fractures of the skull, with only 7 deaths.

Treatment.—The treatment of fractures of the skull has a direct reference either to the existence or to the possible development of the various complication, which have just been referred to. The following general plan will be found most available, if carried out either in part or wholly, according to the necessities of the case.

Control the hemorrhage : remove, at once, all foreign bodies, which can be *readily* reached—such as balls, spiculæ, wadding, dirt, &c. ; wash and bring the edges of the wound gently together; treat symptoms of compression, if present, by placing the patient in the recumbent position with his head lower than his body, using external stimulation, if the pulse fails; when reaction is established, or when there are symptoms of compression or inflammation from the start, bleed and purge freely, use cold applications to the head, enjoin perfect rest, give repose to the special senses, as far as practicable, enforce the lowest diet ; and, finally, when all other means have failed, and symptoms of cerebral compression, inflammation, effusion of blood, or the formation of pus exist to such an extent as to render the diagnosis a matter of no difficulty, resort to the trephine and give the patient the last chance for his life.

It is true that the weight of authority, so far as

writers on modern Surgery are concerned, preponderates against such an employment of this instrument; but, after a due consideration of their arguments and statistics, an attentive study of the works of the older masters, and no little personal observation and experience, the advice in regard to this instrument, is freely given with the full assurance of its reliability and propriety in this connexion. In some instances of compound fracture with depression, or with the penetration of a foreign body into the substance of the brain, the trephine may be immediately employed, but this is not the general rule, as has been previously stated. Neither chloroform nor ether should be used in this operation for fear of inducing inflammation of the brain.

Portions of the skull sliced off by the sabre or sword should be replaced and secured by wire sutures, even if they are attached by small shreds of the scalp.

In all scalp wounds, however caused, avoid the use of sutures, and guard against the development of erysipelas. In simple divisions of the scalp, from blows or cuts, the edges of the wound may be readily kept in apposition by crossing the hairs at different points and binding them by means of small shot.

Hernia cerebri should be treated in its earlier stages by well conducted, systematic compression. Pressure should be made with a piece of sheet lead, a compress and a roller changed as often as may be necessary to ensure firmness and cleanliness. As the mass recedes, the compress is gradually pushed into an osseous opening until it is reduced to the level of the brain.

If by any accident the protrusion has attained to considerable bulk, the proper plan is to exercise all that is accessible or to destroy it with Vienna paste or the actual cautery.

FRACTURE OF THE BONES OF THE FACE.—The bones of the face which present the greatest importance in this connexion are the malar, the nasal and the upper and lower maxillary.

Fractures of Malar Bone.

Causes.—Direct violence—such as blows or falls.

Symptoms.—Depression of the bone, tilting upwards of orbital plate, and protrusion of the eye.

Treatment.—Push the bone into position by carrying the finger through the mouth into the temporal fossa.

Fracture of the Nose.—*Causes.*—Falls or blows.

Symptoms.—Depression of the bone, and interference with nasal breathing,—or lateral deviation of the nose.

Treatment.—Insert the finger or some suitable instrument and elevate the bone. In lateral deviations restore the nose to its proper position, and keep it there by means of adhesive strips.

Fracture of Upper Maxillary Bone.—*Causes.*—Falls, blows and wounds.

Symptoms.—Displacement, deformity, and separation of the bones.

Treatment..—Mould the bones into shape; save every osseous fragment; and keep parts in apposition by means of adhesive plaster.

Fracture of Inferior Maxillary Bone—This bone may be fractured through its body, angle or ramus and condyles.

Causes.—Direct blows, kicks from horses, sword cuts, bullets, &c.

Symptoms.—Fractures of the body are characterized by displacement, mobility, crepitus and pain. The displacement is greater in proportion as the fracture is nearer the symphysis, and less as it approaches the angle.

Salivation and swelling of the sub-maxillary gland, together with difficulty of speech and deglutition are soon developed. A fracture of the ramus may be distinguished by a grating noise at the seat of the injury and great pain about the ear.

A fracture of the neck may be determined by crepitation in moving the jaw, preternatural mobility in front of the ear and the dragging forward of the bone by the external pterygoid muscle. Fracture of the condyle may be readily distinguished in the same way. In connexion with compound fractures of this bone and the bones of the face generally; hemorrhage, paralysis—from division of the branches of the facial nerve—inflammation, and constitutional irritation; produced by swallowing the secretions from the wound, are likely to occur.

Treatment.—In cases of simple fracture, seat the patient upon a chair; support his head upon the breast of an assistant and let it be firmly held; pass the fingers along the base of the jaw, or the fractured portion of the bone; mould the parts into proper shape; close the mouth, take care that the lower teeth rest firmly against the upper; then adapt a piece of paste board or felt wet with hot water to the base and sides of the jaw; and finally apply either Gibson's or Barton's bandages, so as to press the lower jaw firmly against the other. If the bone be comminuted and the teeth forced from their socket, the latter should be returned, and secured to the sound ones by silver wire.

In compound fractures care should be taken to preserve as much of the bone as possible, to keep the fragments in apposition, to arrest hemorrhage by compressing or ligating either the external carotid or the facial artery, to counteract the disturbing agency of muscles by compresses, bandages and adhesive strips, and to see that the secretions from the wound are not swallowed. Inflammation should be treated on general principles. Fluid food must be administered for several weeks when semi-solid nourishment may be substituted.

FRACTURES OF THE BONES OF THE TRUNK.—*Fracture of the Clavicle.*—*Causes.*—Falls, blows, wounds from sabres, bullets, shells, &c.

Varieties.—Fractures of the clavicle may be simple, compound, comminuted, complicated, unilateral, bilateral, transverse or oblique. The usual seat of fracture is at the middle of the bone where it is weakest.

Symptoms.—Sunken appearance of the shoulder; shoulder drawn downwards, inwards and forward by the weight of the limb and the action of the deltoid, subclavius and pectorial muscles; inclination of head and trunk to affected side; impossi bility of rotating the arm by carrying hand to the face; crepitation, elicited by pushing the shoulder upwards, outwards and backwards; separation of the fragments, the outer being drawn downwards inwards and forwards, and the inner fragment, slightly upwards by the sterno-cleido-mastoid muscle.

Treatment.—The great indication is to carry the shoulder upwards, outwards and backwards, until the outer fragment reaches the level of the inner fragment, and to retain it there. The reduction of the fracture may be readily accomplished, but retention is more difficult. The simplest and most effectual method of keeping the fragments in apposition, is to place a pad in the axilla; to bring the elbow against the antero-lateral aspect of the chest and to place the fore-arm against the front; to carry the fingers across the opposite clavicle; and then to apply adhesive strips, reaching around the limb and shoulders, and binding the arm down to the chest. If this is not sufficient the apparatus of Velpeau, Fox, Levis or Dugas may be employed. Compound, comminuted and complicated fractures should be treated on general principles, remembering that the great indication is to carry the shoulder upwards, backwards and outwards and to retain it there until union has taken place.

Fractures of the Scapula.—These are of rare occurrence, especially in civil Surgery. When the *acromion* process is broken the accident generally produced by violence applied to the upper and outer parts of the shoulder.

Symptoms.—The shoulder loses its rotundity; the fractured portion is drawn downwards and forwards by the action of the deltoid muscle; the fragments rests upon the front and upper part of the head of the humerus; and the limb is movable; while the signs of the accident are effaced when the arm is elevated.

Treatment.—The indication is to relax the deltoid muscle by carrying the arm forward across the chest, and by raising the elbow up so that the head of the humerus may press against the acromion process. The same apparatus as for fractured clavicle may be used, dispensing with the axillary pad.

When the *Coracoid* process is broken—which is a rare accident—the fragment is carried inward and downwards, by the conjoined action of the pectoralis major and the coraco-brachialis muscles.

Treatment.—Flex the fore arm and carry the arm forwards across the chest; place a pad in the axilla; and push the humerus upwards. Retain the arm in position by means of adhesive strips.

When the body of the bone is broken, there is no displacement. The bone should be steadied by applying pads and keeping them in position by means of adhesive strips or rollers carried around the chest.

Fractures of the Ribs.—Causes. Violence of all kinds, such as falls, blows, gunshot wounds, and muscular action, &c. The central ribs being more exposed are most frequently broken.

Symptoms.—Displacement of fragments with crepitation such as can be felt with the hands when the patient coughs; a peculiar cracking noise following a deep inspiration; pain at the seat of injury, increased by the respiratory efforts; spitting of blood, together with pleuritic and pneumonic symptoms, dyspnœa and emphysema, if either fragments or spiculæ have been pushed inwards; and copius exteanal hemorrhage when the intercostal artery has been divided. It not unfrequently happens in compound fractures of the ribs that large spiculæ of bone are driven deeply into the parenchyma of the lungs, causing violent inflammation, hemorrhage, escape of air into the cavity of the pleura, disappearance of the respiratory murmur unusual resonance, &c., followed either by speedy death or protracted suppuration. Again, a spent ball may impinge with some violence against a rib, not fracturing the bone but seriously implicating the delicate structures beneath it.

Treatment.—In simple fracture without serious displacement, encircle the chest with a broad bandage or strip of adhesive plaster, so that the intercostal muscles may be put in motion as little as possible in connexion with the respiratory function. If there be outward displacement the same apparatus, with the addition of compresses may be employed. If the displacement be inwards it should not be interfered with, unless complicated by serious symptoms connecting themselves with the lungs or pleura. In such a contiugency, after failing to afford relief to the patient by a proper use of pressure and antiphlogistic remedies, the fragment may be raised by mechanical means. If .there be dangerous hemorrhage from an intercostal artery it may be compressed against a rib or drawn out and tied.

In compound fractures the depressed portions should be elevated; the spiculæ and foreign bodies removed; the pain, cough, &c., incident to pulmonary lesions treated with opium administered in large doses; hemorrhage arrested by copious bleeding from the arm, so as to produce syncope and induce the formation of clots in the divided vessels; the wound closed as soon as hemorrhage ceases; the patient placed upon the wounded side‡ so as to promote adhesion and facilitate the escape of all fluids; digitalis or veratrum administered to control the circulation; inflammation treated on general principles; and pus or air evacuated, if it forms in sufficient quantity to embarrass the circulation seriously. If there be the serious dyspnœa the bandage should not be applied; and in examining the wound, the finger instead of the probe should be employed, lest the delicate tissue of the lung be more seriously irritated. When inflammation of the lung and pleura supervene upon blows which do not fracture the rib, opium should be freely administered, and the symptoms treated on general principles.

In penetrating wounds of the lung the danger is *primarily* from hemorrhage and collaps and *secondarily* from inflammation and its products. Distinct plans of treatment are consequently demanded at different periods in the history of the case:

---

† In wounds from stabs this rule should be rigidly adhered to, but in gun shot wounds the patient may be allowed to assume the position most agreeable to him.

1. The employment of means for arresting the flow of blood—such as venesection, opium, &c.

2. The employment of such remedies as are required to arrest inflammatory reaction, or to guard the system against the deleterious effects of the products of that process.

If venesection be attempted, the patient should be placed in the erect position and a large opening made, so that syncope may be as speedily produced as possible. The question of the propriety of bleeding is one which frequently exercises all the judgment at the command of the Surgeon, for though venesection is *the remedy* when properly employed, it s far from being of universal application. The following circumstances may be regarded as furnishing contra-indications to the employment of the lancet in wounds of the lung.

1. When a considerable time has elapsed after the receipt of the wound, and a large amount of blood has been lost.

2. When the patient is weak and anæmic in consequence of the debilitating influences incident to the regime of camps and hospitals.

3. When the patient has been debilitated by previous disease or wounds.

4. When the large vessels leading to or from the heart are severed.

5. When the patient has received other wounds of a serious character.

6. When the nervous shock incident to the wound is overwhelming.

7. When the erect posture cannot be borne.

8. When proper subsequent treatment is impos-

sible as in hurried marches, hasty retreats, want of the means of transportation, the impossibility of securing reliable and continuous surgical assistance, &c.

9. When the patient is manifestly *in articulo mortis*.

Wounds of the lung are far from being so fatal as might be supposed in advance. Numerous cases have come under my own observation, during the present war, in which rapid recoveries have followed the most severe penetrating wounds of this delicate organ. The experience of Confederate Surgeons will confirm the assertion that unless death speedily results from hemorrhage and collapse a favorable prognosis may be formed in a majority of such cases.

STATISTICS.

Reported by	Macleod,	122 cases.	98 deaths.
"	Legouest,	6 "	3 "
"	Guthrie,	106 "	53 "
"	Meniere,	29 "	9 "

Fracture of the Pelvis.—Causes.—Great violence and gunshot wounds.

Symptoms.—The usual signs of fracture in connexion with some serious complication, such as laceration of bladder or rectum, injury to the peritoneum, division of arteries and veins.

Treatment.—Keep the patient in bed and treat the complication on general principles. When the Os-coccygis is broken, the finger should be introduced into the rectum and the fragments replaced.

FRACTURES OF THE BONES OF THE SUPERIOR EXTREMITIES.—*Fractures of the Humerus.*—The Head, anatomical neck, surgical neck, shaft and condyles of the Humerus may be fractured.

Fracture of the head.—The head of the Humerus is frequently fractured by balls, though this accident from other causes, is very uncommon. If the fracture be compound the fragments can readily be felt with the fingers. The treatment under these circumstances, is resection. Boyer states that there can be no bony union when the fracture is is intra-capsular, and that death is generally the result.

Fracture of the Anatomical neck.—*Causes.*—Falls and blows, a rare accident.—*Symptoms.* The head can be felt in the glenoid cavity; slight hollow below the acromion; axis directed inwards; crepitaion very faint; and the bones shortened slightly.

Treatment.—A pad in the axilla, splints to the shaft, and a sling to keep the elbow slightly raised.

Fracture of the Surgical neck.—*Causes.*—Falls upon the hand; direct violence, and muscular action.

Symptoms.—The upper fragment slightly elevated by the muscles attached to the tuberosities; the upper end of lower fragment drawn inwards by latissimus dorsi, pectoralis major and teres major muscles: humerus thrown obliquely outwards by deltoid muscle, and sometimes elevated so as to project beneath and in front of the coracoid process.

Treatment.—The indications are to counteract the action of the opposing muscles and to keep the fragments in position. Draw the arm from the body; apply four paste board splints on its sides; place a large conical shaped pad with its base upwards, in the axilla; approximate the elbow to the side and retain it there by strips of adhesive plaster or a broad roller passed around the chest; flex the forearm and support it in a sling.

Fracture of the Shaft.—*Causes.*—Falls, violence; muscular contraction, &c.

Symptoms.—Deformity, preternatural mobility and crepitus. There is but little shortening, as the weight of the arm counteracts it. If the fracture be below the deltoid the inferior fragment will be drawn inwards, but if above that point, outwards. The limb is powerless and is supported by the patient at the wrist. The fracture may also be compound, complicated or comminuted, the diagnosis being easy in each case. When the fracture occurs just above the condyles the lower fragment is carried backward and upwards by the action of the triceps.

Treatment.—In simple fracture, apply a roller from the fingers to the axilla; adjust either two, three or four splints, made of paste board, sole leather or thin wood, to the arm,—one extending from the axilla to within an inch of the condyle, another from the shoulder joint to an inch above the corresponding condyle, a third in front and a fourth behind; flex the forearm and support in a sling. Reunion will generally occur in a month. When the fracture is complicated with a division of the artery, ligation in the wound, should be immediately resorted to. When the fracture is compound, the patient should be put to bed, and the injured limb supported upon a pillow, the forearm being kept at an obtuse angle with the arm, the elbow on a level with the shoulder, and the hand little higher than the elbow. No bandage should be applied, but support may be given either by wire splints—permitting irrigation—or two lateral wooden splints. The patient must be kept perfectly quiet, so that the upper fragment may not be disturbed by any movement of the trunk.

When the swelling has subsided, and the inflammation has been subdued,—the starch bandage may be used with advantage.

Fracture of the Condyles.—The causes producing this fracture are the same as those already referred to under previous heads.

Symptoms.—The detached condyle can usually be felt with the finger; crepitus is perceived on bending the arm; there is pain at the seat of injury with deformity. If the inner condyle be fractured, the ulna projects backward, but resumes its natural position when the arm is extended; while the humerus advances in front of the ulna. If the external condyle be separated the joint is immovable, the hand remains supine, and there is constant semiflexion of the forearm.

Treatment.—Coaptate the fragments; retain them in position by means of compresses, and strips of adhesive plaster, applied around the joint in the form of a figure of eight; apply the angular splint; and support the forearm in a sling.

Fractures of the Ulna.—The Olecranon process,

coronoid process and shaft of this bone are liable to be broken.

Fracture of the Olecranon.—*Causes.*—Direct violence, and muscular action.

Symptoms.—Semiflexion of the limb; impossibility of extending the forearm; a hollow at the back of the elbow; a prominence at the posterior inferior surface of the arm; pain, swelling, &c.; crepitus when the radius is rotated.

Treatment.—Bring the separated parts into position; confine them by means of compresses and adhesive strips; apply a wooden splint in front of the joint; and keep the arm extended.

Fracture of the Coronoid process.—*Causes.*—Direct injury, as the passage of the wheel of a coach, or force applied to the hand, impelling the ulna and radius violently upwards against the lower extremity of the humerus.

Symptoms.—The ulna is carried backwards and upwards; the olecranon is prominent; the limb cannot be flexed; the detached bone can be felt above the elbow; crepitation, pain and swelling, present themselves.

Treatment.—Bandage the forearm carefully from the fingers and the upper arm from the shoulder downwards; flex the forearm at a right angle; and enclose the arm in a tin case or angular splints. Adhesive plaster may be carried around the joint so as to keep the fragment in position.

Fracture of the Shaft.—*Causes.*—Direct violence, counter stroke, muscular action, &c.

Symptoms.—A marked depression at the inner border of the forearm, mobility of the fragments, crepitation, pain, swelling, and displacement of the lower fragment.

Treatment.—Apply a long splint in front with a compress adjusted so as to preserve the interosseous space, and another behind,—both extending from the elbow to the end of the fingers, and wider than the arm; have the forearm in a position midway between pronation and supination; let the thumb project as a guide; and then bind the splints to the forearm by means of a roller bandage. Or the hand may be permanently inclined towards the thumb, by means of two splints the extremities of which are made somewhat sloping from behind forwards.

Fracture of the Radius.—The superior extremity, shaft and inferior extremity of this bone may be broken. Fracture of Superior Extremity.—*Causes.* Direct violence.

Symptoms.—Deformity below the joint; projection of the upper end of the lower fragment; impossibility of rotating the forearm; the refusal of the upper fragment to follow the motions of the lower, &c.

Treatment.—Place the limb at right angles with the arm in a position midway between pronation and supination, and employ the same splints as for fracture of both bones of the forearm.

Fracture of the Shaft.—*Causes.*—Violence direct or indirect.

Symptoms.—The fragments approach the interosseous space; while there is more or less of deformity, preternatural mobility, absence of the power to pronate and supinate the army, and crepitus.

Treatment.—Precisely the same as for fracture

of the ulna. When the curved or pistol handle splints are used they should be sloped from before backwards.

The shafts of the Radius and Ulna are frequently fractured together by a direct blow or indirect violence. The fragments are drawn inwards by the pronator quadratus, tending to destroy the interosseous space. Reduction is readily made by extension from the wrist, and retention is effected by means of splints, padded so as to preserve the interosseous space, and extending from the elbow to the end of the fingers. The splints should b wider than the arm, and no attempt should be made to bandage the limb before their application. The arm should be in a position midway between pronation and supination and the thumb left out as a guide. The indications are to preserve the interosseous space and to prevent the upper fragment of the radius from being too much supinated.

Fracture of lower end of the Radius.—The radius may be broken either directly at the joint, or an inch and a half above it. The former is known as Barton's and the latter as Colle's fracture.

Causes. Violence either direct or indirect.

Symptoms.—The lower fragment is drawn upwards and backwards behind the upper fragment, by the combined action of the supinator longus and the flexors and extensors of the thumb and carpus, producing a prominence on the back of the wrist and a deep depression behind. The upper fragment projects forward, and is drawn by the pronator quadratus in close contact with the ulna, causing a projection on the anterior surface of the forearm just above the carpus—all the usual signs of fracture are also present.

Treatment.—The treatment consists in flexing the forearm, and making powerful extension from the wrist and elbow, depressing at the same time the radial side of the hand, and retaining the parts in position by compressing each projecting point and the use of well padded pistol shaped splints. Bond's and Smith's splints are regarded as the best for this fracture.

In compound fractures of the forearm, the patient should be put to bed, and the arm placed upon a pillow or in a well padded fracture box. Cold water should be allowed to drip upon it from above. Care should be taken to keep the arm semipronated, to ensure the parallelism of the bones, and to have the pillow made firm by placing a wide board beneath it. When the swelling and inflammation have subsided, the arm may be placed in a starch bandage or on a wide splint and supported by a sling, when the patient can walk about.

Fracture of the Carpal Bones should be treated on general principles.

Fractures of the Metacarpal Bones.—Causes.—Direct and indirect blows, gunshot wounds, &c.

Symptoms.—One fragment is elevated above the other. The deformity can be readily reduced but again shows itself when the pressure is removed.

Treatment.—Make moderate extension upon the finger corresponding to the broken bone; force the fragments into position; apply paste board splints to the palm, back of the hand and fingers. The splints should be well padded.

Fractures of the fingers can be readily detected.

When the extreme phalanx is broken, the remedy is amputation. When the other phalanges are broken, coaptation may be ensured by extension, and the fragments retained in position by means of a splint made of paste board or felt, moulded accurately to either the dorsal or palmar aspect of the finger. Compound fractures should be treated on general principles. Conservative Surgery holds a proud pre-eminence in this connexion, and the operator should endeavor to save as much of the member as possible.

Fractures of the Femur.—This bone may be broken either in its upper extremity, in its shaft or in its inferior extremity.

Fracture of the neck, internal to the capsular ligament. *Causes.*—Direct violence, indirect violence, such as slipping off the edge of a curbstone, gunshot wounds, &c.

Symptoms.—Slight shortening of the limb; eversion of the foot from the combined action of the external rotator muscles, together with the psoas, iliacus; preternatural mobility—shown by rotating the limb upon its axis, flexing it upon the pelvis, or extending it behind the line of the sound limb; change of position in the great trochanter—being drawn upwards towards the ilium and in close contact with the acetabulum, and also describing a smaller segment when the limb is rotated; change of attitude—the body is thrown forward; the sound limb is firmly planted on the floor, the unsound one hangs off in a constrained and awkward manner—the foot and knee being everted, the leg is supported upon the ball of the toes, while the heel is elevated two or three inches, the natural prominence of the hip is destroyed, and the patient cannot walk. This accident usually occurs in persons of advanced age, because the neck of the bone is more horizontal at that period, and the bone contains more earthly matter. The union between the fragments is always of a fibro-ligamentous nature, when it takes place at all.

Treatment.—The plan recommended by Sir Astley Cooper, and pursued by most surgeons, is to keep the patient quietly in bed for two or three weeks and then permit him to walk about on crutches. The long splint such as will be described in connexion with factures of the shaft, may also be employed. In compound fractures at this point, resection may be attempted under the modifications alluded to in another portion of this volume.

Fracture at the base of the neck, (extra capsular.) *Causes.*—Falls upon the hip; blows; falls upon the foot or knee; gunshot wounds, &c.

Symptoms.—Shortening: eversion of the foot or knee; mobility of the fragments; distinct crepitation; elevation of the trochanter; severe pain; great swelling; considerable shock followed by excessive reaction.

Treatment.—Place the limb in the straight position; apply splints on either side; and make extension and counter extension according to the plans which will be more fully explained when fractures of the shaft are considered. The foot should be inclined slightly outwards to relax the rotator muscles and great care should be taken to prevent over lapping of the fragments or angular deformity. Continue the dressings for at least five weeks.

It is a matter of great moment to distinguish between intra-capsular fracture and iliac dislocation and between fracture within and without the capsular ligament. The following signs will establish the diagnosis.

INTRA-CAPSULAR FRACTURE.	ILIAC DISLOCATION.
1. Occurs generally in old persons and most common in women.	1. Most frequently in adult and middle life—and is common to both sexes.
2. Produced usually by slight causes.	2. Produced by great violence.
3. Foot strongly everted.	3. The foot is inverted.
4. Great shortening, which returns readily after reduction.	4. Shortening does not return after reduction.
5. Crepitation.	5. No crepitation.
6. Preternatural mobility.	6. The bone is fixed and in a constrained position.

INTRA-CAPSULAR FRACTURE.	EXTRA-CAPSULAR FRACTURES.
1. Slight shortening which gradually increases to two inches and upwards.	1. Shortening is less but more persistent.
2. Crepitation indistinct.	2. Crepitation very distinct.
3. Function impaired.	3. Loss of function complete.
4. Trochanter moves on rotation, as it were, upon a pivot.	4. The trochanter is only partially separated and imperfectly obeys the movements of the limb.
5. Pain greatest in the direction of the small trochanter.	5. Pain severe and located near the great trochanter.
6. But slight swelling, contusion or dislocation.	6. Severe contusion with considerable swelling, ecchymosis and discoloration.

In compound fractures from gunshot wounds resection may possibly be resorted to, but the wisest plan is to attempt to save the limb. After much observation and reflection, I am convinced that the probabilities of a favorable issue, are much increased by the rejection of all appliances in the way of inclined planes, extending and counter extending forces, &c. They tend to increase irritation and inflammation, to interfere with water dressings, and the free discharge of pus, and to render the patient more uncomfortable, while they do not secure better results so far as the usefulness and symmetry of the limb are concerned. Smith's anterior splint may be tried in the premises, but if the case does not progress favorably, all dressings should be removed, and the limb placed on such a position upon pillows as will best secure the comfort of the patient. The preservation of the sufferer's life is the great desideratum, while the usefulness and symmetry of the member are matters of secondary consideration in this connexion.

Fracture of the shaft in its upper third. The most common seat of this fracture is from two and a half to three inches below the trochanter minor.

Causes.—Direct or indirect violence, &c.

Symptoms.—The upper fragment is carried forwards by the action of the psoas and iliacus internus, and at the same time everted and drawn outwards by the external rotator and glutei muscles, causing a marked prominence at the outer side of the thigh and great pain from the laceration of the muscles; the lower fragment is drawn upwards, by the rectus, biceps, semi-membranosus and semi-tendinous muscles, whilst its upper end is thrown outwards and its lower end inwards by the pectineus and adductor muscles; crepitation, preternatural mobility and the ordinary signs of fracture, are also present.

Treatment.—This fracture may be treated, when simple, either by direct relaxation of all the opposing muscles by means of the double inclined plane, or by overcoming the contraction of the muscles by the use of the long splints. Of these two plans of treatment preference should be given to that for relaxing the muscles, which can be most successfully accomplished by the doubled inclined plane. Mode of procedure. Obtain if possible a proper bed; apply a roller from the toes to the groin; secure two splints made of binder's boards, softened in hot water and nearly meeting in front, to the thigh; lay the limb over the double inclined plane, which should be well cushioned; attach the foot to the foot board, so as to prevent inversion or eversion of the member; raise the body slightly so as to relax the psoas and iliacus muscles; adjust the angle beneath the knee in such a manner as to relax the muscles by which the lower fragment is kept out of position, and to keep the two fragments upon the same plane, and in the same line; bind the limb to the apparatus by means of a roller bandage; and retain it in position either by means of pegs placed on the side, or by side boards so arranged as to form with the splint a kind of trough. By means of Smith's anterior splint the conjoint advantages of relaxation and extension may be secured. This apparatus is nothing more than a double enclined plane made of strong iron wire as long as the limb and applied anteriorly, with cords passing from the upper and lower cross wires and uniting into a common one, which passes in an oblique direction to the wall and suspends the limb. The muscles are not only relaxed, as by the ordinary inclined plane, but all tendency to contraction is obviated by the obliquity of the cord, which acts as the extending force below, and by the weight of the body, which serves as the counter extending force above. The suspension of the limb precludes such displacements as are likely to occur in consequence of the movements of the body, thus securing a much greater latitude in that regard and contributing materially to the comfort of the patient.

If there be trouble in keeping the upper fragment in place, a compress or an additional splint may be applied above it, so as to force it in position. Care must be taken not to apply the bandage too tightly, or to permit the bed clothes to rest upon the limb.

In compound fractures of the upper third of the thigh from gunshot wounds, the same principles will apply as enumerated above. The long splint tends to augment both the local and constitutional irritation, while the double enclined plane, and even Smith's apparatus, soon become irksome to the sufferer and tend to interfere with the free escape of pus, by causing it to gravitate towards the body. The better plan is to reject them in the first instance, to place the limb in a comfortable position upon a pillow, to resort at once and persistently to the cold water treatment, and to direct every energy towards the preservation of the patient's life, reserving the question of deformity for a later period in the history of the case. An attempt should be made to save the limb on account of the extreme fatality of amputations in this locality.

Fracture of the middle third. *Causes.*—Same as last.

Symptoms.—The superior fragment overlaps the inferior; the lower end of the superior fragment is drawn inwards and upwards by the flexor muscles; the limb is shortened from 2 to 4 inches, and everted; the upper end of the inferior fragment forms a projection on the forepart of the thigh; while mobility, crepitus, pain and swelling contribute their quota to the perfection of the diagnosis.

Treatment.—Numerous plans have been devised for the treatment of this fracture, but the following seems to possess the greatest advantages. Directions: lay the perineal band in its place and place four pieces of bandage transversely where the broken thigh is to rest; over these lay a splint as wide as the diameter of the thigh, well padded, and long enough to reach from the tuberosity of the ischium to the lower margin of the ham; lay the patient upon the bed, with his thigh reposing upon the back splint and his head and body slightly raised; make an assistent seize the knee firmly and make moderate traction, so as to steady the limb; lay long strips of adhesive plaster upon the leg from the knee down, forming loops below, and and secured to the limb by other strips and a roller carried spirally around it, taking care to protect the ankles by small pieces of cotton batting; apply then a roller from the toes to the ham; lay the long splint on the *outside* of the limb, extending from four to five inches below the foot either to the crest of the ilium, according to Desault, or to the axilla as suggested by Physick; adjust the perineal band, and attach the upper extremity of the long splint to the body by means of a band passed around it; twist the adhesive strips below the foot into a small rope, attach them to the extending screw in the foot piece, and tighten them moderately so that the assistant may release his hold upon the knee; lay a padded splint upon the *inside* of the limb extending from the groin to a point immediately below the knee; apply another splint in *front* extending from the groin to within one inch of the knee; bring up the four transverse bands, previously placed under the limb, so as to include the three short splints and the long splint; then carry extension to the utmost point of tolerance; fill up all the inequalities and insterstices with soft cotton; and complete the dressing by applying a roller bandage over the splints from the foot to the groin. Increase the extension daily for a week, and then maintain it until union is complete. About the twenty eighth day relax the extension, and lift the limb regularly, rubbing and gently flexing the knee. For two months the patient should walk on crutches and bear but little weight upon the limb.

In compound fractures, especially where there is much comminution, but little advantage can be expected from extension with the long splint until the violence of the inflammatory action has subsided; while the inclined plane is liable to the objections which have already been referred to. Gentle extension may be attempted by applying and securing adhesive strips to the leg, then attaching a weight to them and suspending it over a pulley at the foot of the bed; but if the case does not progress favorably even this should be discarded and the fracture treated upon the plan suggested in connexion with similar injuries in the upper hird. For statistical information in regard to the treatment of compound fractures of the femur, see table "I," of the Appendix. Malgaigne declares that in the attempt to save the limb, under these circumstances, no greater risk is run than in amputating it. This is certainly an extreme view of the case, as is established by the statistics of Stone and Baudens,—who themselves are advocates of conservative surgery—in connexion with these accidents. The femur has frequently been resected for injuries of this character, but the expediency of this procedure is very questionable.

Fracture of the Lower third—immediately above the condyles. *Causes.* Direct or indirect violence.

Symptoms.—The lower fragment may be felt in the popliteal space being drawn back by the gastrocnemius, soleus and plantaris muscles, and upwards by the rectus; the end of the upper fragment is drawn inwards by the pectineous and adductor muscles, and forwards by the psoas and iliacus; the limb is shortened; while crepitation, pain, swelling, &c., are also present.

Treatment.—The indication is to relax the opposing muscles, and approximate the broken fragments. This is accomplished by placing the limb upon the double inclined plane.

The principal circumstances which demand the attention of the Surgeon in connexion with fractures of the shaft of the femur are:

1. The ends of the broken bone must be steadily kept upon the same plane and in a line with each other.

2. Care must be taken that no shortening occurs. The extending and counter extending bands should be watched and tightened when necessary.

3. The limb should be placed in a slightly *elevated* position.

4. It is important to keep as much pressure off the heel as possible in order to prevent sloughing, and ulceration.

5. The perineal band must be carefully watched and care taken to prevent it from excoriating the parts beneath.

6. The bandage should not be applied too speedily, tightly or irregularly, and in compound fractures should be dispensed with. The starch bandage may frequently be employed to great advantage.

7. The bed cloths must be kept off the fractured limb, lest they disturb the fragments.

8. Passive motion of the neighbouring joints should be undertaken at the end of the twenty eighth day.

Fracture of the Patella.—*Causes.*—Direct injury, as a fall or blow; indirect violence; and muscular contraction.

Symptoms.—In transverse fracture, the upper

fragment is displaced; the aspect of the limb is changed; the limb cannot be extended; there is some pain, but no crepitation.

Treatment.—Extend the leg; elevate the foot; bring the fragments together and retain them in apposition by means of adhesive strips; and place the limb upon an inclined plane.

Fracture of the Tibia.—The shaft of the tibia is most frequently broken obliquely at the lower fourth of the bone, by direct or indirect violence.

Symptoms.—If the fracture has taken place from above downwards and forwards the fragments ride over one another, the lower fragment being drawn backwards and upwards by the muscles of the calf; while the pointed extremity of the upper fragment projects forwards beneath or through the integument. If the direction of the fracture is the reverse of this, the pointed extremity of the lower fragment projects forwards, riding over the lower end of the upper one. There is but little crepitation, and not much pain. The internal malleolus is most frequently broken off about the centre of that process, in an oblique direction.

Treatment.—Bend the knee so as to relax the muscles; bring the fragment in apposition; apply adhesive straps from the point of fracture, and form a loop below the foot; tie a cord to this loop, with a weight attached to it, and pass it over a small wheel at the foot of the bed; when inflammation has subsided; apply the starch bandage and cut off the straps close to the foot. In compound fractures, the same plan may be adopted, taking care to leave the wound uncovered. The fracture box, filled with bran may also be employed. An admirable plan is to suspend the leg in a sling reaching from the knee to the foot. The wound should be treated on general principles.

Fractures of the Fibula.—Fractures of the head and shaft of the bone are so readily detected and easily treated that no particular discription is necessary. When the fracture occurs in the inferior fifth of the bone the accident is a more serious one. *Causes.* Forcible abduction of the foot, such as occurs in falls; and direct violence.

Symptoms.—When the fibula alone is broken, there will appear slight eversion of the foot; depression at the seat of injury; and change in the aspect of the joint. When the malleolus is broken off, or when the tibia has given way a short distance above the articulation, the foot seems to be dislocated outwardly; the malleoli are widely separated; a deep pression in the line of fracture presents itself; the foot is unusually movable, while its external margin is elevated and its internal depressed; crepitation can be heard; while there is considerable pain, swelling and ecchymosis.

Treatment.—The indication is to maintain the foot in a position the reverse of that which is caused by the injury. This is accomplished by Dupuytren's apparatus, which consists of a light wooden splint and a wedged shaped cushion,—the former reaching from the upper third of the leg to about three inches below the sole of the foot, and the latter from the same point to a level with the ankle. Bandage the limb, but do not compress it opposite the site of fracture; stretch the apparatus along its inner surface, with the tapering end of the pad upwards; and secure it first above and then below, carrying the roller around the foot and ankle in such a manner as to turn the internal margin of the foot upwards and inwards. The limb may then be kept extended, or half bent upon a pillow. Attempt passive motion at the end of a week.

Both the Tibia and Fibula may be broken contemporaneously. If the fracture be transverse, there is no danger of deformity; but if oblique there will be considerable shortening. In oblique fractures, therefore extension and counter extension must be made and persisted in until union has taken place, while the simple fracture box will answer for transverse fractures, provided the great toe is kept constantly on a line with the inner border of the patella.

Fractures of the Foot.—The calcaneum may be broken by direct violence. There is always considerable contusion and laceration of the soft parts. The signs by which this accident may be determined are a hollow at the heel; a protuberance at the lower and back part of the leg; and the impossibility of extending the foot. The fragments should be brought together and a complete relaxation of the muscles of the calf secured, by keeping the leg in a permanently extended condition. In fractures of the other bones there is no displacement.

Experience teaches that in gunshot wounds of the foot involving a fracture of its bones, there is always danger to be apprehended, however seemingly insignificant the injury. The bones from their peculiar conformation, are easily shattered, each fragment becoming the focus of an extensive inflammation, which speedily produces pus in large quanties. In consequence of the thickness of the fasciae, the purulent matter does not readily escape, but burrows in every direction, causing intense pain, and great nervous irritation, and inducing pyaemia with all its frightful consequences.

If the ball does not pass entirely through the foot, it should be immediately sought for, and a counter opening made, if possible, to facilitate the discharge of pus. The endermic exhibition of morphia may also be resorted to for the purpose of relieving the pain incident to the wound, of preventing the development of tetanic symptoms, and of securing quietude and sleep to the patient. The foot should be kept in an elevated position until the development of pus, but not longer, and the wound treated on general principles.

TABLE A.
SPECIAL OPERATIONS.

APPENDIX.

PERIOD.	HEY'S OPERATIONS.			CHOPART'S OPERATIONS.			PIROGOFF'S OPERATIONS			SYME'S OPERATIONS.			TREPHINE'S OPERATIONS.		
	No. of Operations.	Conv.	Died.	No. of Operations.	Conv.	Died.	No. of Operations.	Conv.	Died.	No. of Operations.	Conv.	Died.	No. of Operations.	Conv.	Died.
Primary,	1	1											2		2
Secondary,				1	1		1	1		2	1	1			
Intermediary,													1	1	
Aggregate,	1	1		1	1		1	1		2	1	1	3	1	2

B.

CONSOLIDATED TABLE

Of Capital Operations performed in and around Richmond, Va., from June 1st to August 1st, 1862, in C. S. Hosp'l.

AMPUTATION OF LEG.

PERIOD IN WHICH OPERATED UPON.	SPECIFYING OPERATION AND ITS SEAT.												Total No. of operations.			(Grand total of operations.)	Died.			Total of deaths.	Recovered or Convalescent.			Total recovered or convalescent.	Per centage of Deaths.			Total per centage of Deaths.
	Upper third.			Middle third.			Lower third.			Not stated.																		
	Circular.	Flap.	Not stated.	Circular.	Flap.	Not stated.	Circular.	Flap.	Not stated.	Circular.	Flap.	Not stated.	Circular.	Flap.	Not stated.		Circular.	Flap.	Not stated.		Circular.	Flap.	Not stated.		Circular.	Flap.	Not stated.	
Primary.	9	2	30	3	3	2			2			18	16	5	47	72	8	2	19	29	12	3	8	43	50	22	40[4]	41
No. of deaths.	5	3	12	2	2	1			1			8	8	2	19	29	2	2	8	12				18	33	50	40	40
Intermediary.	1	1	12	1	1	4		1				4	6	4	20	30	3				1	4	12	13	75	100	45[4]	56[6]
No. of deaths.	5	4	7	4	1	1	1					14	2	4	8	12												
Secondary.	2	4	8	4	3	1	1					7	4	4	22	30		2		17				74	50	47	44	43[9]
No. of deaths.	2	9		5	2							7	3	4	10	17												
Aggregate.	16	49	42[6]	60	75	50	25	4	16			36	26	17	89	132	13	5	37	58	17	5	52	74				
Ratio of deaths.	58	56		60	75							41	30	47	41[4]	43[9]												

C.

CONSOLIDATED REPORT.

Of Capital Operations performed in and around Richmond, Va. from June 1st, to Aug. 1st, 1862, in C. S. Hosp'ls.

AMPUTATION OF THIGH.

PERIOD IN WHICH OPERATED UPON.	SPECIFYING OPERATION AND ITS SEAT.											Total No. of Operations.			Died				Recovered or Convalescent.				Percentage of Deaths.			
	Upper third.			Middle third.			Lower third.			Not stated.																
	Circular.	Flap.	Not stated.	Circular.	Flap.	Not stated.	Circular.	Flap.	Not stated.	Circular.	Flap.	Not stated.	Circular.	Flap.	Not stated.	Grand total of Operations.	Circular.	Flap.	Not stated.	Total of Deaths.	Circular.	Flap.	Not stated.	Total recovered or convalescent.	Circular.	Flap.	Not stated.	Total per cent. of deaths.
Primary, No. of deaths.	1		4	4	2	7	5	2	10	2		23	16	10	44	70	5	3	18	26	7	7	30	44		30	318	36⁷
Intermediary, No. of deaths.	1	1	3	2	1	7	2	2	11			6	9	8	14	26	6	5	31	45	8	1	3	12	63	83	524	90
Secondary, No. of deaths.	1		2	2	2	11	2	2	11			20	5	6	16	44	5	1	21	28	4	1	8	13	422	50	74₈	63
Aggregate, Ratio of deaths.	2¹	1	7¹⁷	3¹¹	4¹⁷	21¹⁷	9¹⁷	5¹⁷	32¹⁷	2		67⁴	32⁵	18	122	172	18	9	76	103	12	8	45	69	561	50	624	52⁸

D.
CONSOLIDATED REPORT.

Of Capital Operations performed in and around Richmond, Va., from June 1st, to Aug. 1st, 1862, in C.S. Hosp'l.

AMPUTATION OF FORE-ARM.

PERIOD IN WHICH OPERATED UPON.	SPECIFYING SEAT OF OPERAT'N.				Total no. of Operations.	Died.	Remaining under treatment.	Recovered.	Per centum of Deaths.
	Upper third.	Middle third.	Lower third.	Not stated.					
Primary.	7	10	6		23	2		21	8[5]
No. of deaths,	1	1			2				
Intermediary,		8	5		13	2		11	15[3]
No. of deaths,		1	1		2				
Secondary,	1	5	3		9	2		7	22[2]
No. of deaths,		1	1		2				
Aggregate,	8[5]	23	14		45	6		39	13[3]
Ratio of deaths.		13	14[2]		13[3]				

APPENDIX.

CONSOLIDATED REPORT.

Of Capital Operations performed in and around Richmond, Va., from June 1st, to Aug. 1st, 1862 in C. S. Hosp'ls.

AMPUTATION OF ARM.

PERIOD IN WHICH OPERATED UPON.	SPECIFYING SEAT OF OPERAT'N.				Total no. of operations.	Died.	Remaining under treatment.	Recovered.	Per centum of deaths.
	Upper third.	Middle third.	Lower third.	Not stated					
Primary,	22	57	13		92	16		76	17
No. of deaths,	1	11	4		16				
Intermediary,	16	31	2		49	22		27	44
No. of deaths,	11	11			22				
Secondary,	10	38	3		51	16		35	31
No. of deaths,	3	11	2		16				
Aggregate,	48	126	18		192	54		138	28
Ratio of deaths.	31	26	33						

F.

CONSOLIDATED REPORT.

Of Disarticulations performed in and around Richmond, Va., from June 1st, to October 1st, 1862.

PERIOD IN WHICH OPERATED UPON.	Shoulder joint.	Elbow joint.	Wrist joint.	Knee joint.	Ankle joint.	Total No. of Operations.	Died.	Recovered or Convalescent.	Per centage of Deaths.
Primary, No. of deaths.	14	—	1	—	—	15	9	6	60
Intermediary, No. of deaths.	9	1	—	—	2	12	7	5	58*
Secondary, No. of deaths.	3	1	1	2	1	8	5	3	60
Aggregate, Ratio of deaths.	26	2	2	2	3	35	21	14	60

G.

CONSOLIDATED TABLE.

Of Capital Operations performed in and around Richmond, Va., from June 1st. to Aug. 1st, 1862, in C. S. Hosp'ls.

RESECTIONS.

PERIOD IN WHICH OPERATED UPON.	Clavicle.	Shoulder joint.	Elbow joint.	Hip joint.	Knee joint.	Inf. maxillary.	Total No. of Operations.	Died.	Recovered or Convalescent.	Per centum of Deaths.
Primary,	1	3	1	—	1	—	5	2	3	40
No. of Deaths.	—	1	1	—	—	—	2			
Intermediary,	—	1	2	—	—	2	5	2	3	40
No. of Deaths.	—	1	1	—	—	—	2			
Secondary,	—	1	3	1	—	1	7	2	5	28³¹
No. of Deaths.	—	—	2	—	—	—	2			
Aggregate,	1	5	6	1	1	3	17	6	11	35⁴
Ratio of deaths.	—	40	66⁶	—	—	—	35³			

H.

CONSOLIDATED TABLE.

Of Ligations of Arteries in Richmond, Va., from June 1st, to August 1st, 1862, in C. S. Hospitals.

PERIOD AT WHICH PERFORMED.	Carotid artery.	Subclavian artery.	Axillary artery.	Brachial artery.	Femoral artery.	Total No. of Operations.	Died.	Recovered or Convalescent.	Per centum of Deaths.
Primary,	1	1	1		5	6	5	1	83.3
No. of deaths.	1	1	1		4	5	4	1	80
Intermediary,	2			1	2	5			
No. of deaths.	1			1	1	4	2	4	33.4
Secondary,	3	1	1	2	3	6	6	6	64.7
No. of deaths.	1				1	2			
Aggregate,	3				10	17			
Ratio of deaths.	66.4	100	50	60	64.7				

I.

TABLE OF COMPOUND FRACTURES OF THIGH,

From gunshot wounds treated without amputation in Confederate Hospitals of Richmond and its vicinity, from May 1st, to October 1st, 1862.

SEAT OF INJURY.				MODE OF TREATMENT.										RESULT.			
Upper third.	Middle third.	Lower third.	Not stated.	Double incl'd plane.	Per cent. of deaths.	Smith's ant. splint.	Per cent. mort.	Liston's splint.	Per cent. mort.	Straight splint.	Per cent. mort.	Not stated.	Per cent. mort.	Recovered.	Died.	Average shortening.	Per cent. of deaths.
83	57	33	38	59	67.79	46	46.66	3	100	16	37.50	78	62.89	80	121	1¾	60.19

This calculation is based upon the reports furnished from the Richmond Hospitals, and gives far more favorable results than have been obtained elsewhere. It is probable that multitudes of unfavorable cases which died early, are not embraced in this table. The per centage of mortality throughout the Confederacy has been much higher than that recorded above.

www.ingramcontent.com/pod-product-compliance
Lightning Source LLC
Chambersburg PA
CBHW020101020526
44112CB00032B/799